A programmed text in statistics

BOOK 4
Tests on variance and regression

Other Books by G. B. Wetherill

Elementary Statistical Methods.
Sequential Methods in Statistics.
Sampling Inspection and Quality Control.

A programmed text
in statistics

BOOK 4
Tests on variance and regression

J. HINE
Formerly Research Officer
University of Bath

G. B. WETHERILL
Professor of Statistics
University of Kent, Canterbury

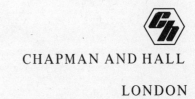

CHAPMAN AND HALL

LONDON

First published 1975
by Chapman and Hall Ltd.,
11 New Fetter Lane, London EC4P 4EE

Typeset by
E.W.C. Wilkins Ltd.
London & Northampton
Printed in Great Britain by
Whitstable Litho

ISBN 0 412 13750 X

© 1975 J. Hine and G. B. Wetherill

This paperback edition is sold subject to the condition that it shall not, by way of trade or otherwise, be lent, re-sold, hired out, or otherwise circulated without the publisher's prior consent in any form of binding or cover other than that in which it is published and without a similar condition including this condition being imposed on the subsequent purchaser.

All rights reserved, No part of this book may be reprinted, or reproduced or utilized in any form or by any electronic, mechanical or other means, now known or hereafter invented, including photocopying and recording, or in any information storage and retrieval system, without permission in writing from the Publisher.

Distributed in the U.S.A.
by Halsted Press, a Division
of John Wiley & Sons, Inc., New York

Library of Congress Catalog Number 75–1230

Contents

	page
Preface	vii
Section 1	1
Tests concerning variances	1–5
Summary	5
One tailed and two tailed tests	6–11
Summary	11
The F-distribution	11–15
Steps in application [6]	
Summary	16
Exercises for Section 1	17
Physical sciences and engineering	17

BOOK 4. TESTS ON VARIANCE AND REGRESSION

Corrigenda

Page
22	Line 13	*for* the sign of the constant A; *read* The sign of the constant B.
29	Line 3	under 'regression coefficient', *for* the sign of the constant A; *read* the sign of the constant B.
30	Frame 1	In the second response, *for* A; *read* B.
31	Frame 6	In the first response, *for* 0.46; *read* -0.46.

Coefficient of correlation	32–41
Summary	41

Exercises for Section 2   42

    Physical sciences and engineering   42

    Biological sciences   43

    Social sciences   45

Solutions to Exercises, Section 1   47

    Physical sciences and engineering   47

    Biological sciences   49

    Social sciences   49

Solutions to Exercises, Section 2   51

    Physical sciences and engineering   51

    Biological sciences   55

    Social sciences   58

Tables   62

    χ^2 — tests involving variances   62

    χ^2 — one tailed tests   63, 64

    χ^2 — two tailed tests   65

    F-distribution   66–69

Preface

This project started some years ago when the Nuffield Foundation kindly gave a grant for writing a programmed text to use with service courses in statistics. The work carried out by Mrs. Joan Hine and Professor G. B. Wetherill at Bath University, together with some other help from time to time by colleagues at Bath University and elsewhere. Testing was done at various colleges and universities, and some helpful comments were received, but we particularly mention King Edwards School, Bath, who provided some sixth formers as 'guinea pigs' for the first testing, the Bishop Lonsdale College of Education, and Bradford University.

Our objectives in the text are to take students to the use of standard t, F and χ^2 tests, with some introduction to regression, assuming no knowledge of statistics to start, and to do this in such a way that students attain some degree of understanding of the underlying reasoning.

Service courses are often something of a problem to statistics departments. The classes are frequently large, and the students themselves are very varied in their background of mathematics. Usually, a totally inadequate amount of time is allocated for an extensive syllabus. The solution offered here is to use the programmed text in place of lectures, and have a weekly practical class at which students work through the exercises given at the end of each section. The available staff effort can then be placed in helping individuals out of particular difficulties, rather than in mass lecturing.

At one time we envisaged producing about three short video taped lectures to discuss some of the important concepts: random variation, probability distribution and sampling distribution. Unfortunately finance could not be obtained for this, and attempts to do it 'on the cheap' proved impossible, although it is to be hoped that this will be achieved in the future.

The main text units are written using examples drawn from many fields of application. In addition, several different sets of exercises are provided, each dealing with a particular field, so that students can apply the techniques within their own sphere of interest. The alternative sets of exercises can either be ignored, or used as 'spares'. Exercises should be written in a notebook, and submitted regularly. Further, students should make their own summaries in a notebook, using as a basis the 'summaries' given at the end of each section.

Those using the text on their own should train themselves to complete all exercises relevant to their particular field of interest, and make notes as described above.

We have endeavoured to keep the mathematical level of the text low; however, it is inevitable that some concepts and notation have to be included. We would particularly mention the use of the exponential (e)

in Book 2 section 2. It would be impractical to describe this function in the text but anyone not familiar with it can either consult their tutor or read about it in any basic mathematics book. Other notation is adequately described in the text.

The production of this text in four units is designed for several reasons. Firstly, it avoids giving the student a volume of such a size that it would be discouraging. Secondly, if some lecturers prefer not to use certain sections, they can easily replace these by standard lectures, or write their own text units. Thirdly, it facilitates use of the text for slightly different syllabuses. Books 3 and 4 can be used individually to study the use of t, F, and to give a simple introduction to regression. However, if this is done, the material presented in Books 1 and 2 must be covered in some other way.

It may be helpful here to explain something about the method of construction of the programmed texts. Once the objectives and entry behaviour are stated, the material to be covered is analysed and put into a 'flow chart'. Material not relevant is omitted. We have done all this for a basic common core of material, and clearly some students may require something more. In particular, some groups of students may require rather more on probability, and a coverage of statistical independence. We have prepared an alternative version of the earlier units, for those with greater mathematical ability, but this is not yet available commercially. Programmed texts on probability alone, however, are available from several sources.

Our thanks are due to Mrs. J. Honebon for her painstaking work in typing the manuscripts, and retyping them many times for the trials.

J.H.
G.B.W.
August 1974

HOW TO USE THE TEXT

Although there are prose passages, the bulk of the text is in programmed form. A statement is given on the left, and this includes gaps in which you have to judge the correct response. The column on the right gives the answer, and this should be covered until you have decided what response you are making.

NOTE: Books 3 and 4 are conveniently used together even if the material in Books 1 and 2 is covered in some other way. Book 4 could also be used on its own. Before studying Book 4, students must have covered the ideas of significance tests and confidence intervals for a normal mean when σ is known. It would also be natural to have covered the t-test.

Section 1

TESTS CONCERNING VARIANCES

Introduction

We have discussed methods of testing the means of samples and populations using the Normal and t-distributions and we have also discussed the use of the χ^2 distribution in testing for goodness of fit and independence of two criteria in contingency tables.

But, we have not yet considered any problems involving the *variances* of samples and population.

For example, we have not yet considered any problems of the following type:-

A random sample of 10 batteries have a standard deviation equal to 1.2 years. Test the hypothesis that σ, the population standard deviation, is equal to 0.9 year.

This test, in fact, makes use of the χ^2 distribution.

Use of χ^2 distribution

In the sections GOODNESS OF FIT and CONTINGENCY TABLES we calculated a value of χ^2 using the equation

$$\chi^2 = \sum \frac{(O-E)^2}{E}$$

and then tested using the χ^2 distribution with $(n-1)$ degrees of freedom.

However, it can be shown mathematically that the expression $\dfrac{(n-1)s^2}{\sigma^2}$ where s^2 = sample variance, σ^2 = population variance, n = sample size

has a χ^2 distribution on $(n-1)$ degrees of freedom.

Thus, in order to test the hypothesis that **σ** is equal to a given value, we compute a value of **χ²** using

$$\chi^2 = \frac{(n-1)s^2}{\sigma^2}$$

and test using the **χ²** distribution with $(n-1)$ degrees of freedom.

However, the χ^2 test for variances differs slightly from that for goodness of fit.

The table we use in testing for goodness of fit gives, for varying degrees of freedom, values of χ^2 for which A = 0.05 and A = 0.01 (see Fig. 1.1)

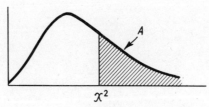

Fig. 1.1.

and in testing for goodness of fit we only consider a χ^2 value significant if it is *greater* than these 5 per cent or 1 per cent significance levels — thus indicating an unduly large difference between the observed and expected values.

Such a test is called a ONE-TAILED TEST.

Now, in testing the hypothesis that σ is equal to a given value, we have to take into consideration the fact that σ could be *significantly greater* OR *significantly less* than this value and so in tests involving variances it is apparent that we cannot use the χ^2 table used in goodness of fit tests.

The table of χ^2 used is given on page 62. A χ^2 value is considered to be significant if it lies *outside* these significance levels.

Such a test is called a TWO-TAILED TEST.

We will now work the example given in the introduction.

1. Firstly, as in all examples involving tests of hypothesis, we set up the ... ... i.e. for this example, we assume	null hypothesis $\sigma = 0.9$
2. Thus we have $\sigma = ...\ ;\ s = ...\ ;\ n = ...$ (numerical values) and so using $\chi^2 = \dfrac{(n-1)s^2}{\sigma^2}$ we can calculate our χ^2 value i.e. $\chi^2 = \dfrac{......}{...}$ =	0.9 1.2 10 $\dfrac{9 \times (1.2)^2}{(0.9)^2}$ 16

3.

We now test using ... with ... | χ^2 distribution with 9 d.f.

Using the table given on page 62 we see that for this distribution
1 per cent sig. level: $\chi^2 = ...$ and $\chi^2 = ...$ | 1.735 and 23.589
5 per cent sig. level: $\chi^2 = ...$ and $\chi^2 = ...$ | 2.700 and 19.023

4.

The value $\chi^2 = 16$ is ... | not significant
at 5 per cent level.
Therefore, there is ... that σ is not equal to 0.9 and we | no evidence
... the null hypothesis that $\sigma = 0.9$; | accept

5.

Thus, in general terms, the four steps in testing the hypothesis that σ is equal to a given value σ_0 are

STEP 1. Set up ... ... | null hypothesis
        i.e. assume, ... | $\sigma = \sigma_0$

STEP 2. Calculate a χ^2 value using

$$\chi^2 = \frac{...}{...}$$ | $\chi^2 = \dfrac{(n-1)s^2}{\sigma^2}$

where
     $s^2 = ...$ | sample variance
     $\sigma^2 = ...$ | population variance
     $n = ...$ | sample size

STEP 3. Test using, ... | χ^2 distribution with $(n-1)$ d.f.

STEP 4. Draw conclusions.

    Significance at 1 per cent level — ... evidence that $\sigma \neq \sigma_0$ | almost conclusive
    ∴ ... null hypothesis. | reject

    Significance at 5 per cent level — ... evidence that $\sigma \neq \sigma_0$ | reasonable
    ∴ ... null hypothesis. | reject

    No significance — ... evidence that $\sigma \neq \sigma_0$ | no
    ∴ ... null hypothesis. | accept

6.

A random sample of 8 cigarettes of a certain brand has an average nicotine content of 18.6 milligrammes and a standard deviation of 2.4 milligrammes. Test the hypothesis that $\sigma = 1.5$.
    Give full working.

If you had any difficulties with this question go to Frame 7. | Significance at 5 per cent level
If not go to Frame 9. | $\therefore$ reasonable evidence that
 | $\sigma \neq 1.5$ and we reject the null
 | hypothesis.

7.

STEP 1. Null hypothesis: ...   Assume $\sigma = 1.5$

STEP 2. Calculate χ^2 using

$$\chi^2 = \frac{(n-1)s^2}{\sigma^2}$$

where $s^2 = ...$ sample variance
    $\sigma^2 = ...$ population variance
    $n = ...$ sample size
In this example, $\sigma = ...,$ 1.5
    $s = ..., n = ...$ 2.4 8

$$\therefore \chi^2 = \frac{......}{.} \qquad\qquad \frac{7 \times (2.4)^2}{(1.5)^2}$$

$$= \frac{......}{...} \qquad\qquad \frac{7 \times 5.76}{2.25}$$

$$\therefore \chi^2 = 18.$$

STEP 3. We now test using ..., ... χ^2 distribution with 7 d.f.
  For this distribution, 1 per cent significance level
    $\chi^2 = ...$ and $\chi^2 = ...$ 0.989 and 20.278
  5 per cent significance level
    $\chi^2 = ...$ and $\chi^2 = ...$ 1.690 and 16.013

  The value $\chi^2 = 18$ is ... at ... level.   significant 5 per cent

STEP 4. Draw conclusions.
  Therefore ... evidence that $\sigma \neq 1.5$ and we   reasonable
  ... the null hypothesis.   reject

8.

A random sample of size 19 from a normal distribution has a mean $\bar{X} = 32.8$ and a standard deviation $s = 4.51$. Test the hypothesis that $\sigma = 9$.

Give full working.

A complete solution is given below.

9.

A production process gives components whose strengths are normally distributed with mean 40 lb and standard deviation 1.15 lb.

A modification is made to the process which may alter the standard deviation of the strength measurements.

The strength of nine components selected from the modified process gave a standard deviation of 1.32 lb.

Is there evidence to assume that the modification has altered the standard deviation of the strength measurements?

Give full working.

A complete solution is given on p. 6.

Summary

To test the hypothesis that $\sigma = \sigma_0$

STEP 1. Set up null hypothesis.
Assume $\sigma = \sigma_0$.

STEP 2. Calculate a χ^2 value using $\chi^2 = \dfrac{(n-1)s^2}{\sigma^2}$

where s^2 = sample variance
σ^2 = population variance
n = sample size.

STEP 3. Test using χ^2 distribution with $(n-1)$ d.f.

STEP 4. Draw conclusions.

Significance at 1% level — almost conclusive evidence that $\sigma \neq \sigma_0$
Therefore reject null hypothesis.

Significance at 5% level — reasonable evidence that $\sigma \neq \sigma_0$
Therefore reject null hypothesis.

No significance — no evidence that $\sigma \neq \sigma_0$
Therefore accept null hypothesis.

Solution to problem given in Frame 8

Null hypothesis: Assume $\sigma = 9$

Calculate χ^2: $\sigma = 9$, $s = 4.51$, $n = 19$

$$\therefore \chi^2 = \frac{18 \times (4.51)^2}{(9)^2}$$

$$= 4.52.$$

Test using χ^2 distribution with 18 d.f.
1 per cent significance level $\chi^2 = 6.265$ and $\chi^2 = 37.156$
5 per cent significance level $\chi^2 = 8.231$ and $\chi^2 = 31.526$
The value $\chi^2 = 4.52$ is significant at 1 per cent level.
Therefore almost conclusive evidence that $\sigma \neq 9$ and we reject the null hypothesis.

Solution to problem given in Frame 9

This problem in fact reduces to testing the hypothesis that $\sigma = 1.15$

Calculate χ^2: $\quad \sigma = 1.15, \; s = 1.32, \; n = 9$

$$\therefore \chi^2 = \frac{8 \times (1.32)^2}{(1.15)^2} = 2 \frac{\cancel{8} \times \cancel{1.74}\;0.58}{\cancel{1.32}\;\cancel{0.33}\;0.11}$$

$$\chi^2 = 10.56.$$

Test using χ^2 distribution with 8 d.f.
1% significance level $\chi^2 = 1.344$ and $\chi^2 = 21.955$
5% significance level $\chi^2 = 2.180$ and $\chi^2 = 17.535$
The value $\chi^2 = 10.56$ is not significant at 5% level.
Therefore no evidence that $\sigma \neq 1.15$ and we accept the null hypothesis.
We conclude that there is no evidence that the modification has altered the standard deviation of the strength measurements.

ONE- AND TWO-TAILED TESTS

Introduction

In all the examples we have been asked to 'test the hypothesis that $\sigma = \sigma_0$' and to do this we have tested using the χ^2 distribution whether or not σ is significantly greater than OR significantly less than σ_0, i.e. we have carried out a TWO-TAILED TEST. In fact we have tested the null hypothesis that $\sigma = \sigma_0$ against the hypothesis that $\sigma \neq \sigma_0$.

But now consider the following two examples.

EXAMPLE 1
Suppose that in the past the average length of life of street lamp bulbs has been 1608 burning hours with a standard deviation of 300 hours. A check sample of bulbs delivered one month is examined and 8 bulbs selected gave lengths of life (in hours) with standard deviation 500 hours.
   Is this evidence that the standard deviation for the month's bulbs is in excess of 300 hours?

EXAMPLE 2
Sixteen rounds of ammunition have been fired, giving a standard deviation of barrel pressures equal to 1.3 thousand pounds/square inch. Based on this information is it reasonable to conclude that the standard deviation of such barrel pressures is less than 2.0 thousand pounds/square inch?
   The tests of hypotheses in these examples involve ONE-TAILED TESTS. We still test the null hypotheses that $\sigma = \sigma_0$, but instead of testing against the hypothesis that $\sigma \neq \sigma_0$, we test, in example 1, against the hypothesis that $\sigma > 300$ and in example 2, against the hypothesis that $\sigma < 2.0$.
The hypothesis *against* which we test the null hypothesis is called the ALTERNATIVE HYPOTHESIS.
i.e. in example 1, the alternative hypothesis is that σ is significantly greater than 300 and in example 2, the alternative hypothesis is that σ is significantly less than 2.0.

Thus, the calculations are basically the same as for two-tailed tests except in that the significance levels used are different.

In two-tailed tests the 1 per cent (5 per cent) significance levels are the values of the χ^2 distribution *outside* which we can be 1 per cent (5 per cent) sure that our value of χ^2 will lie.

In one-tailed tests, to test the alternative hypothesis that $\sigma > \sigma_0$, the 1 per cent (5 per cent) significance level is the value of χ^2 *above* which we can be 1 per cent (5 per cent) sure that our value of χ^2 will lie and

in one-tailed tests, to test the alternative hypothesis that $\sigma < \sigma_0$, the 1 per cent (5 per cent) significance level is the level *below* which we can be 1 per cent (5 per cent) sure that our value of χ^2 will lie.

The tables of significance levels for one- and two-tailed tests are given on pages 63, 64 and 65.
In order to clarify these ideas we will work the two examples given.

Example 1

1.

We first set up the hypotheses:-
Null hypothesis: ... $\sigma = 300$
Alternative hypothesis: ... . $\sigma > 300$

2.

Secondly we calculate a χ^2 value using ... (in symbols).
$$\chi^2 = \frac{(n-1)s^2}{\sigma^2}$$

In this example,
$n = ..., s = ..., \sigma = ...$ (by assumption) $\quad n = 8, s = 500, \sigma = 300$

$$\therefore \chi^2 = \frac{...}{...} \quad \text{numerical values.}$$
$$= ...$$

$\dfrac{7 \times 500 \times 500}{300 \times 300}$

19.4

3.

We now find the level of significance using the ... distribution with ... d.f.

χ^2
7 i.e. $(n-1)$

Since we are testing against the alternative hypothesis that σ is greater than 300 we use the tables given on page 64.

Thus, 1 per cent significance level is
$$\chi^2 = ...$$ 18.475
5 per cent significance level is
$$\chi^2 = ...$$ 14.067

4.

The value $\chi^2 = 19.4$ is significant at the ... level. | 1%
Thus we (accept/reject) the null hypothesis and | reject
(accept/reject) the alternative hypothesis. | accept

i.e. we have almost conclusive evidence that σ ... ... | is greater than 300

Example 2

Sixteen rounds of ammunition have been fired, giving a standard deviation of barrel pressures equal to 1.3 lbf/in². Based on this information is it reasonable to conclude that the standard deviation of such barrel pressures is less than 2.0 lbf/in² ?

5.

State the null and alternative hypotheses.

Null Hypothesis:
$\sigma = 2.0$ lbf/in³.

Alternative Hypothesis:
$\sigma < 2.0$ lbf/in².

6.

Calculate χ^2.

$$\chi^2 = \frac{15 \times 1.3 \times 1.3}{2.0 \times 2.0}$$

$$= 6.34$$

7.

We now find the level of significance using the ... distribution with ... d.f. | χ^2
 | 15

Since we are testing against the alternative hypothesis that σ is *less* than 2.0 we use the tables given on p. 63.

Thus, 1 per cent significance level is
$$\chi^2 = ...$$ 5.229
5 per cent significance level is
$$\chi^2 = ...$$ 7.261

8.

The value $\chi^2 = 6.34$ (is/is not) significant at the 5 per cent level.

is, since $\chi^2 = 6.34$ lies outside the 5 per cent level but not outside 1 per cent level.

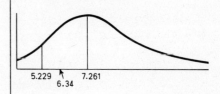

Thus, we (accept/reject) the null hypothesis and (accept/reject) the alternative hypothesis.
i.e. we have reasonable evidence that σ, ...

reject
accept
is less than 2.0 lbf/in².

9.

State the null and alternative hypotheses in the following three examples.

I. 26 schoolchildren between the ages of $9\frac{1}{2}$ and 10 years were given extra pasteurised milk for 4 months and their change in weight (gain or loss) measured, the standard deviation of these measurements being 30 oz. A very large group of schoolchildren of similar age who were not given milk had a standard deviation of change in weight of 25 oz. Test whether those children given the extra milk had a more variable gain in weight over the four months or not.

Null hypothesis: ...
Alternative hypothesis: ... .

$\sigma = 25$
$\sigma > 25$

II. The intelligence quotients of 31 grammar school boys have a standard deviation equal to 7. Test whether this standard deviation is significantly different from 10.

Null hypothesis: ...
Alternative hypothesis: ...

$\sigma = 10$
$\sigma \neq 10$

III. Test the hypothesis that the standard deviation in weight of boxes of a particular brand of cereal is less than 0.03 oz if the weights of a random sample of 10 boxes had a standard deviation equal to 0.01 oz.

Null hypothesis: ...
Alternative hypothesis: ...

$\sigma = 0.03$
$\sigma < 0.03$

Now work the three examples given above.
(Complete solutions are given on p. 10)

Solutions to examples given in Frame 9

Example I

Null hypothesis: $\sigma = 25$
Alternative hypothesis: $\sigma > 25$

$$n = 26, \quad s = 30, \quad \sigma = 25.$$

$$\text{Therefore } \chi^2 = \frac{25 \times 30 \times 30}{25 \times 25}$$

$$= 36.$$

Now test using χ^2 distribution with 25 d.f. and one-tailed test.

1% significance level $\chi^2 = 44.314$ (using tables on p. 64)
5% significance level $\chi^2 = 37.652$.

The value $\chi^2 = 36$ is not significant at the 5 per cent level and we accept the null hypothesis.
i.e. we conclude that there is no evidence that the children given the extra milk had a more variable gain in weight.

Example II

Null hypothesis: $\sigma = 10$
Alternative hypothesis: $\sigma \neq 10$

$$n = 31, \quad s = 7, \quad \sigma = 10$$

$$\chi^2 = \frac{30 \times 7 \times 7}{10 \times 10} = 14.7$$

Now test using χ^2 distribution with 30 d.f. and two-tailed test.

1% significance level: $\chi^2 = 13.787$ and $\chi^2 = 53.672$ (using tables on p. 62.)
5% significance level: $\chi^2 = 16.791$ and $\chi^2 = 46.979$

The value $\chi^2 = 14.7$ is significant at the 5% level and we reject the null hypothesis.
i.e. we conclude that there is reasonable evidence that $\sigma \neq 10$.

Example III

Null hypothesis: $\sigma = 0.03$
Alternative hypothesis: $\sigma < 0.03$

$$n = 10, \quad s = 0.01, \quad \sigma = 0.03$$

$$\chi^2 = \frac{9 \times 0.01 \times 0.01}{0.03 \times 0.03} = 1.$$

Now test using χ^2 distribution with 9 d.f. and one-tailed test.

1% significance level: $\chi^2 = 2.088$
5% significance level: $\chi^2 = 3.325$ (using tables on page 63).

The value $\chi^2 = 1$ is significant at the 1 per cent level and we reject the null hypothesis.
i.e. we conclude that there is almost conclusive evidence that $\sigma < 0.03$.

Summary

STEP 1. Set up null and alternative hypotheses –
   Two-tailed test: Null hypothesis: $\sigma = \sigma_0$
                      Alternative hypothesis: $\sigma \neq \sigma_0$
   One-tailed test: Null hypothesis: $\sigma = \sigma_0$
                      Alternative hypothesis: $\sigma < \sigma_0$
   or   Null hypothesis: $\sigma = \sigma_0$
                      Alternative hypothesis: $\sigma > \sigma_0$

STEP 2. Calculate a χ^2 value using $\chi^2 = \dfrac{(n-1)s^2}{\sigma^2}$

   where s^2 = sample variance
         σ^2 = population variance
         n = sample size.

STEP 3. Test using a χ^2 distribution with $(n-1)$ d.f. and the appropriate significance levels from the tables on pages 62–65.

STEP 4. Draw conclusions.

   Significance at 1% level — almost conclusive evidence to accept alternative hypothesis and reject null hypothesis.
   Significance at 5% level — reasonable evidence to accept alternative hypothesis and reject null hypothesis.
   No significance   — no evidence to accept alternative hypothesis. Therefore accept null hypothesis.

THE F-DISTRIBUTION

Introduction

In the next section we will outline the method of comparing two variances or standard deviations. This test makes use of the *F-DISTRIBUTION*.

Suppose we have two samples taken from two different populations,

Sample 1: size n_1, variance s_1^2, population variance σ_1^2

Sample 2: size n_2, variance s_2^2, $(< s_1^2)$, population variance σ_2^2

and we wish to test the hypothesis that $\sigma_1^2 = \sigma_2^2$.

We first set up the null hypothesis that $\sigma_1^2 = \sigma_2^2$.

A value of F is then calculated using the expression

$$F = \frac{s_1^2/\sigma_1^2}{s_2^2/\sigma_2^2} \quad \text{where} \quad s_1^2 > s_2^2$$

which reduces to $F = \dfrac{s_1^2}{s_2^2}$ since we have assumed that $\sigma_1^2 = \sigma_2^2$.

We then test for the significance of this value using the F-distribution with $(n_1 - 1)$ and $(n_2 - 1)$ degrees of freedom i.e. the F distribution with the number of d.f. for numerator and the number of d.f. for denominator. This is the general outline of the test used for comparing two population variances; through working an example we will discuss the test in greater detail.

Example.
At the beginning of a course, the members of class A (20 students) and of class B (25 students) were given a comprehensive examination of prior work in the field.
The results were as follows:-

	Mean	Variance	Size
Class A.	80	100	20
Class B.	75	144	25

Test the hypothesis that the population variances are equal.

1.

As in all problems involving tests of hypothesis, we first ... ...
...
i.e. in this example, ...
...

| set up the null hypothesis
| we assume $\sigma_1^2 = \sigma_2^2$

2.

We now have to calculate an F value.
The equation for F has been quoted as

$$F = \frac{s_1^2}{s_2^2} \quad \text{where} \quad s_1^2 > s_2^2$$

A simpler way of writing this is

$$F = \frac{\text{larger sample variance}}{\text{smaller sample variance}},$$

∴ In this example $F = \dfrac{\text{variance of class ...}}{\text{variance of class ...}}$ $\dfrac{B}{A}$

$\phantom{\therefore \text{In this example } F} = \dfrac{...}{...}$ (numerical values) $\dfrac{144}{100}$

$\phantom{\therefore \text{In this example } F} = \ ....$ 1.44

3.

In the introduction we stated that to test the significance of an F value we use the F-distribution with

number of d.f. for numerator

and *number of d.f. for denominator*

In our example, number of d.f. for numerator $= 25 - 1$ $= 24.$	size of class B
Similarly, number of d.f. for denominator $= \ldots - \ldots$ $= \ldots .$	size of class A $20 - 1$ 19
$\therefore$ In order to test for the significance of the value $F = 1.44$ we use the F distribution with ... and ... d.f.	24 and 19 d.f.

4.

The tables of F values are given on pages 66–69.
You will notice there are two tables;
   one giving the 1 per cent significance level
         (p. 66 and 67)
and one giving the 5 per cent significance level
         (p. 68 and 69).

For the F-distribution with 24 and 19 d.f. the 1 per cent significance level is $F = 2.92$ and the 5 per cent significance level is $F = \ldots$	2.11

5.

The value $F = 1.44$ is not significant at the 5 per cent level. $\therefore$ There is ... of a real difference between σ_1^2 and σ_2^2 and we ... the null hypothesis that $\sigma_1^2 = \sigma_2^2$.	no evidence accept

6.

Thus, given the following data

	size	variance
Sample 1	n_1	s_1^2
Sample 2	n_2	s_2^2 ($<s_1^2$)

the four steps in testing the hypothesis that the population variances, σ_1^2 and σ_2^2, are equal, are

STEP 1. Set up null hypothesis i.e.	assume $\sigma_1^2 = \sigma_2^2$

STEP 2. Calculate a value of F using

$$F = \frac{\cdots}{\cdots}$$

$\dfrac{s_1^2}{s_2^2}$ or $\left(\dfrac{\text{larger sample variance}}{\text{smaller sample variance}}\right)$

STEP 3. Find level of significance using F distribution with

$\cdots\cdots$

and $\cdots\cdots$

(How many d.f. ?).

no. of d.f. for numerator
(i.e. $n_1 - 1$)
no. of d.f. for denominator
(i.e. $n_2 - 1$)

STEP 4. Draw conclusions.
Significance at 1 per cent level — $\cdots$ evidence of real difference between σ_1^2 and σ_2^2 $\therefore$ $\cdots$ null hypothesis.
Significance at 5 per cent level — $\cdots\cdots\cdots\cdots$
$\therefore \cdots\cdots\cdots$

No significance — $\cdots\cdots\cdots$
$\therefore \cdots\cdots\cdots$

almost conclusive
reject
reasonable evidence of
real difference between
σ_1^2 and σ_2^2 $\therefore$ reject null
hypothesis.

no evidence of real difference
$\therefore$ accept null hypothesis.

7.

The following table gives data on the hardness of wood stored indoors and outdoors. Test to see whether the variability of hardness is affected by weathering.

	Outside	Inside
Sample size	25	15
Standard deviation	9.2	17

Give full working.

If you had any difficulties with this problem go to Frame 8. If not go to Frame 10.

8.

STEP 1. Set up the null hypothesis i.e. assume...

variability of hardness is not affected by weathering.
i.e. that $\sigma_1^2 = \sigma_2^2$

STEP 2. Calculate a value of F using

$F = \dfrac{\cdots}{\cdots}$ (in words)

$= \dfrac{\cdots}{\cdots}$ (numerical values)

$\simeq 3.4$.

$\dfrac{\text{larger sample variance}}{\text{smaller sample variance}}$

$\dfrac{(17)^2}{(9.2)^2} = \dfrac{289}{84.6}$

14

STEP 3. Find level of significance using, ...	F-distribution with $(15-1)$ d.f. and $(25-1)$ d.f.
You will notice that the significance levels are not tabulated for 14 and 24 d.f. We therefore have to approximate using the values which are tabulated.	
The 1 per cent significance level for 12 and 24 d.f. is $F = ...$ and for 15 and 24 d.f. is $F = ...$	3.03 2.89
Obviously the 1 per cent significance level for 14 and 24 d.f. must lie somewhere between the above two limits.	
Thus we can say that the value $F = 3.4$ (is/is not) significant at the 1 per cent level.	is
STEP 4. Draw conclusions. There is ... evidence of a real difference between σ_1^2 and σ_2^2 and we ... the null hypothesis.	almost conclusive reject

9.

Briefly state the four steps involved in testing the hypothesis that $\sigma_1^2 = \sigma_2^2$.

    STEP 1. Set up null hypothesis.

    STEP 2. Calculate a value of F.

    STEP 3. Find level of significance.

    STEP 4. Draw conclusions.

10.

A standard placement test in mathematics was given to 25 boys and 16 girls. The boys made an average grade of 82 with standard deviation 8 while the girls made an average grade of 78 with standard deviation 7. Test the hypothesis that $\sigma_1^2 = \sigma_2^2$

              Give full working.

A complete solution is given on p. 16.

Summary

	Size	Standard Deviation
Sample 1	n_1	s_1
Sample 2	n_2	$s_2\,(<s_1)$

The four steps involved in *testing the hypothesis that* $\sigma_1^2 = \sigma_2^2$ are

STEP 1. Set up null hypothesis i.e. assume $\sigma_1^2 = \sigma_2^2$.

STEP 2. Calculate a value of F using

$$F = \frac{s_1^2}{s_2^2} \text{ or } \frac{\text{larger sample variance}}{\text{smaller sample variance}}.$$

STEP 3. Find level of significance using F distribution with number of d.f. of numerator (i.e. $n_1 - 1$) and number of d.f. of denominator (i.e. $n_2 - 1$).

STEP 4. Draw conclusions.

Significance at 1% level — almost conclusive evidence of real difference between σ_1^2 and σ_2^2
∴ reject null hypothesis.

Significance at 5% level — reasonable evidence of real difference between σ_1^2 and σ_2^2
∴ reject null hypothesis

No significance — no evidence of real difference between σ_1^2 and σ_2^2
∴ accept null hypothesis.

Solution to problem given in Frame 10.

Sample	Size	Standard Deviation
Boys	25	8
Girls	16	7

Null hypothesis: Assume $\sigma_1^2 = \sigma_2^2$

Calculate F.

$$F = \frac{\text{larger sample variance}}{\text{smaller sample variance}}$$

$$= \frac{(8)^2}{(7)^2}$$

$$= 1.31.$$

Find level of significance using F-distribution with 24 and 15 d.f.

1% significance level — $F = 3.29$
5% significance level — $F = 2.29$.

The value $F = 1.31$ is not significant at 5% level.

Therefore there is no evidence of a real difference between σ_1^2 and σ_2^2 and we accept the null hypothesis.

Exercises for Section 1

PHYSICAL SCIENCES AND ENGINEERING

1. Test the null hypothesis that $\sigma = 0.01$ in for the diameters of certain bolts if in a random sample of size 12 the diameters of the bolt had a variance $s^2 = 0.0004$.

2. Samples of size 25 and 20 taken from two normal populations gave $s_1 = 10$, $s_2 = 18$ respectively. Test the hypothesis that $\sigma_1 = \sigma_2$.

3. Past data indicate that the variance of measurements made on sheet metal stampings by experienced quality control inspectors is 0.16 in^2. Such measurements made by an inexperienced inspector could have too large a variance — perhaps because of inability to read instruments properly — or too small a variance — perhaps because unusually high or low measurements are discarded. If a new inspector measures 25 stampings with a variance 0.11 in^2, test whether the inspector is making satisfactory measurements (i.e. test if the variance of measurements made by the inexperienced inspector is less than 0.16 in^2.)

4. Two surveyors make repeated careful measurements of the same angle. Their results differ only in the figure for seconds of angles. The variances of the two sets of results are

$$s_A^2 = 18.7, \quad s_B^2 = 4.5$$
$$n_A = 6, \quad n_B = 8.$$

Should these results be accepted as evidence that B is the more precise observer?

BIOLOGICAL SCIENCES

1. Test the null hypothesis that $\sigma = 0.01$ in for the diameters of certain bolts if in a random sample of size 12 the diameters of bolts had a variance $s^2 = 0.0004$.

2. Samples of size 25 and 20 taken from two normal populations gave $s_1 = 10$, $s_2 = 18$. Test the hypothesis that $\sigma_1 = \sigma_2$.

3. A fixed dose of a certain drug is known to cause in adults an average increase of 10 in pulse rate with a standard deviation 4. You have found in 20 teenage patients an average increase of 15 with standard deviation 5. Test the hypothesis that the standard deviation of increase in pulse rate for teenagers is greater than 4.

4. Two batches of a certain chemical were delivered to a factory. For each batch ten determinations were made of the percentage of manganese in the chemical, giving $s_1^2 = 0.0862$ and $s_2^2 = 0.0672$. Justify the assumption that $\sigma_1^2 = \sigma_2^2$.

SOCIAL SCIENCES

1. Test the null hypothesis that $\sigma = 0.01$ in for the diameters of certain bolts if in a random sample of size 12 the diameters of the bolts had a variance $s^2 = 0.0004$.

2. Samples of size 25 and 20 taken from two normal populations gave $s_1 = 10$, $s_2 = 18$. Test the hypothesis that $\sigma_1 = \sigma_2$.

3. You have introduced into a process for manufacturing tuning forks an innovation which you hope will yield forks having a smaller variance about the true pitch. The standard deviation produced by the standard process is 1 Hz. and in a sample of 20 forks produced by the new process the sample standard deviation is 0.7 Hz.
Test the hypothesis that, for the new process, the standard deviation is 1 against the alternative hypothesis that it is smaller.

4. A study was made to determine if the subject matter in a physics course is better understood when a lab constitutes part of the course. Students were allowed to choose between a 3-semester-hour course without labs and a 4-semester-hour course with labs.
In the section with labs 11 students made an average grade of 85 with a standard deviation of 3 and in the section without labs 16 students made an average grade of 79 with a standard deviation of 6. Test the hypothesis that $\sigma_1^2 = \sigma_2^2$.

Section 2

REGRESSION

Introduction

So far we have concerned ourselves with sets of data in which only one reading on each of a given set of objects has been recorded e.g. number of peas/pod, I.Q.'s, children's weights, etc.

In this section we will consider sets of data in which there are readings on more than one variable. The following are examples of such sets of data.

I. In an experiment of the capacity of electrolytic cells, four cells were taken and filled with 0.50, 0.55, 0.60 and 0.65 ml of electrolyte respectively. Each cell was charged up at constant current to estimate the capacity and this was repeated ten times on each cell.

The results were as follows:-

Quantity of electrolyte (ml)	0.50	0.55	0.60	0.65
Mean capacity	131.27	133.00	134.55	137.01

II.

Number of pupils per teacher	17	19	12	10	16	13	18	20	22	13	19	8
Percentage of pupils passing 3 GCE subjects	5	6	8	7	5	7	3	4	5	9	3	11

We can plot such sets of data on a graph called a SCATTER DIAGRAM. The scatter diagrams for the two examples given are shown on p. 20.

It can immediately be seen that in each example the points are closely clustered around a straight line. This indicates that a relationship may exist between the variables in each example i.e. between quantity of electrolyte and mean capacity, and between the number of pupils/teacher and the percentage of pupils passing three GCE subjects. The line around which the points cluster is called a REGRESSION LINE. Suppose we have two variables x and y; using a regression line we can, given any value of the variable x,

I.

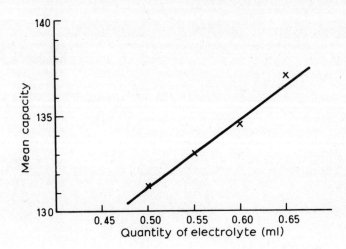

II.

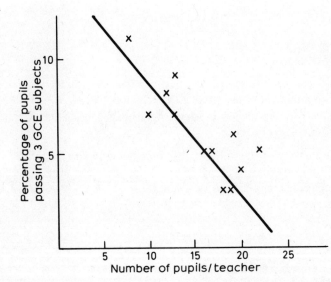

estimate the corresponding value of the variable *y*. For example, in example I, using a regression line we could estimate mean capacity given that there is 0.54 ml (say) of electrolyte. In this case the line is called the regression line of mean capacity on quantity of electrolyte.

In general if a line is used to *estimate a variable y from x*, it is called the *regression line of y on x*, the variable *y* being called the DEPENDENT VARIABLE and the variable *x* the REGRESSOR VARIABLE.

In examples on regression we always plot the scatter diagram before doing any calculations since the scatter diagram shows whether or not a relation is likely to exist. The scatter diagrams given above show that there is a relationship both between mean capacity and quantity of electrolyte and between percentage of pupils passing 3 GCE subjects and number of pupils/teacher. Whereas if we obtained a scatter diagram such as the one given on p. 21 there would be little point in calculating a regression line as it is apparent that no relationship exists between the variables.

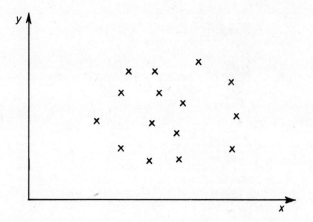

In the examples given on p. 19 we draw the regression lines merely by inspection. The object of regression is to find the 'best' line mathematically.

For a set of n x and y values the equation for this line is found in the following way. Firstly we take any line and find the difference between each of the observed values of y and the corresponding value on the line (see diagram below). These differences are squared and summed. The 'best' line is then the line for which this sum is a minimum and is given by the equation

$$y = \bar{y} + b(x - \bar{x})$$

where $$b = \frac{\Sigma xy - \frac{\Sigma x \Sigma y}{n}}{\Sigma x^2 - \frac{(\Sigma x)^2}{n}}$$

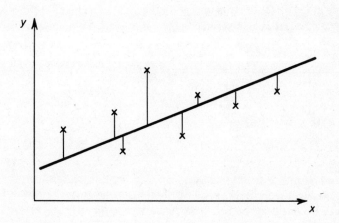

But how do we decide on which variable to take as the *y* variable and which to take as the *x* variable?

In example I, it is apparent that the quantity of electrolyte is under the control of the experimenter whereas the mean capacity is not. Therefore, it is obvious that we need to estimate the mean capacity from the quantity of electrolyte. Thus we take quantity of electrolyte as the *x* variable and mean capacity as the *y* variable. The variable *y* contains the random variation.

Similarly the number of pupils per teacher is *to some extent* under the control of the experimenter. Thus we take this variable as the *x* variable and the percentage of pupils passing 3 GCE subjects as the *y* variable.

Before commencing any examples on regression it is essential that the student should note the following points on the theory of straight lines

1. The equation of a straight line is of the form

$$y = A + Bx$$

2. The sign of the constant *A* indicates the direction of slope of the line.
   If B is positive the line slopes 'upwards' (see Diagram (i))

(i)

If B is negative the line slopes 'downwards' (see Diagram (ii))

(ii)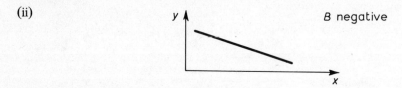

Thus in the regression equation the constant *b* indicates the slope of the line.

We will now work some regression examples.

I. A study was made on the effect of temperature on the yield of a chemical process. The following data (in coded form) was obtained. Calculate the regression of yield on temperature.

Temperature	Yield
−5	1
−4	5
−3	4
−2	7
−1	10
0	8
1	9
2	13
3	14
4	13
5	18

1.

Since we are asked to find the regression of yield on temperature we take yield as the (x/y) variable and temperature as the (x/y) variable.

    y   x

2.

Plot the scatter diagram and draw in the line that you consider to be the 'best' regression line.

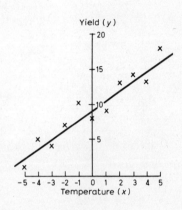

you will notice that the line slopes upwards.
Therefore, the value of b which we calculate should be (positive/negative).
This gives us a rough check on our calculations of b.

positive

3.

The regression equation is

$$y = \bar{y} + b(x - \bar{x})$$

where

$$b = \frac{\Sigma xy - \frac{\Sigma x \Sigma y}{n}}{\Sigma x^2 - (\Sigma x)^2/n}$$

(see introduction).
Thus we first need to calculate the values Σx Σy Σx^2 Σxy.

Complete Table 2.1, p. 24, and hence find the above quantities.

Table 2.1

x	y	x^2	xy
−5	1	25	−5
−4	5	16	−20
−3	4	9	−12
−2	7		
−1	10		
0	8		
1	9		
2	13		
3	14	9	42
4	13	16	52
5	18	25	90
$\Sigma x = $ ...	$\Sigma y = $ ...	$\Sigma x^2 = $ ...	$\Sigma xy = $ ...

4	−14
1	−10
0	0
1	9
4	26

$\Sigma x = 0$, $\Sigma y = 102$,
$\Sigma x^2 = 110$, $\Sigma xy = 158$

4.

In this example we have 11 pairs of values of x and y.
Therefore $n = $ ...

11

In *symbol* form

$$b = \frac{\text{... ... ...}}{\text{... ... ...}}$$

$$b = \frac{\Sigma xy - \dfrac{\Sigma x \Sigma y}{n}}{\Sigma x^2 - \dfrac{(\Sigma x)^2}{n}}$$

Thus, substituting numerical values gives

$$b = \frac{... - ... \times .../...}{... - \dfrac{(...)^2}{n}}$$

$$b = \frac{158 - 0 \times 102/11}{110 - \dfrac{(0)^2}{11}}$$

Therefore $b = $ ... (to 2 d.pl.).

$= 1.44$ (to 2 d.pl.)

5.

Notice that the value of b is positive which (agrees/does not agree) with the fact that the 'rough' regression line slopes upwards.

agrees

6.

We now have to find $\bar{x}$ and $\bar{y}$

$$\bar{x} = \frac{\Sigma x}{n} \quad \bar{y} = \frac{\Sigma y}{n}$$

∴ In this example,

$$\bar{x} = \frac{...}{...}$$

$\dfrac{0}{11}$

$= ...$

$= 0$

$$\bar{y} = \frac{...}{...}$$

$\dfrac{102}{11}$

$= ...$.

$= 9.27$ (to 2 d.pl.)

7.

The regression of y on x is given by ... (in symbols).

$$y = \bar{y} + b(x - \bar{x})$$

Therefore, in this example

$$y = ...(x - ...) + ...$$

$$y = 1.44(x - 0) + 9.27$$

which gives

$$y = ... + ...$$

$$y = 9.27 + 1.44x$$

II. We will now work example II given in the introduction.

No. of pupils per teacher	Percentage of pupils passing 3 GCE subjects
17	5
19	6
12	8
10	7
16	5
13	7
18	3
20	4
22	5
13	9
19	3
8	11

Calculate the regression of pupils passing on number of pupils per teacher.

8.

Since we are asked to find the regression of pupils passing on number of pupils per teacher we take percentage of pupils passing as the (x/y) variable and number of pupils per teacher as the (x/y) variable.

y

x

9.

Plot the scatter diagram and draw in the line you consider to be the 'best' regression line

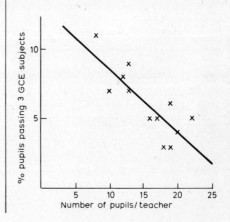

| | You will notice that the line slopes (upwards/downwards) Therefore the value of b should be, ... | downwards negative |

10.

It is apparent that the calculations in this example will be fairly laborious. (Consider, for example, the calculation of x^2.)
We can, however, simplify the working considerably by transforming the variables.

Go to Frame 11.

Table 2.2

x	y	$u = x - 16$	$v = y - 5$	u^2	uv
17	5	1	0	1	0
19	6	3	1	9	3
12	8	−4	3	16	−12
10	7	−6	2	36	−12
16	5	0	0	0	0
13	7				
18	3				
20	4				
22	5				
13	9	−3	4	9	−12
19	3	3	−2	9	−6
8	11	−8	6	64	−48
		$\Sigma u = ...$ ...	$\Sigma v = ...$ ...	$\Sigma u^2 = ...$ ...	$\Sigma uv = ...$ ...

−3	2	9	−6
2	−2	4	−4
4	−1	16	−4
6	0	36	0

19 − 24 = −5   18 − 5 = 13   209 104   3 − 104 = −101

11.

We will now transform the data given in Table 2.2.
First consider the variable x.
We will transform x into a new variable u by subtracting 16 from each value.

Therefore $u = ... − ...$
Complete column 3, Table 2.2.

$x - 16$

12.

Now consider variable y.
We will transform y into a new variable by subtracting 5 from each value.
Therefore $v = ... − ...$
Complete column 4, Table 2.2.

$y - 5$

13.

Transforming the variables makes no real difference to the calculation of b. The same formula applies except in that transformed values are used.

i.e. instead of Σx we use Σu
 instead of Σy we use Σv
 instead of Σxy we use ... Σuv
 instead of Σx^2 we use ... Σu^2

Thus the formula for b becomes

$$b = \ldots .$$ $$\dfrac{\Sigma uv - \Sigma u \Sigma v / n}{\Sigma u^2 - (\Sigma u)^2 / n}$$

14.

Now complete columns 5 and 6, Table 2.2 and use the formula given in Frame 13 to calculate b.

$$b = \dfrac{(-101) - (-5)(13)/12}{209 - (-5)^2/12}$$

$$= -0.46 \text{ (to 2 d.pl.)}$$

15.

Notice that the value of b is negative which agrees with the fact that the 'rough' regression line slopes, ...

downwards

16.

If we transform a variable x to a variable y using the transformation

$$y = x - A$$

then the mean $\bar{x}$ is given by

$$\bar{x} = \bar{y} + A$$

(see Section 2)

Thus, in this example, since

$$u = x - 16$$
$$\bar{x} = \bar{u} + 16$$

and similarly since

$$v = y - 5$$
$$\bar{y} = \ldots$$ $\bar{v} + 5$

17.

Using Table 2.2 calculate $\bar{u}$ and $\bar{v}$ and hence $\bar{x}$ and $\bar{y}$.

$$\bar{u} = \dfrac{-5}{12} \qquad \bar{v} = \dfrac{13}{12}$$

$$= -0.42 \qquad = 1.08$$

$$\therefore \bar{x} = 15.58 \quad \bar{y} = 6.08$$

18.

The regression of y on x is given by

$$y = \ldots \qquad\qquad b(x - \bar{x}) + \bar{y}$$

Thus substituting the numerical values of $b, \bar{x}$ and $\bar{y}$ gives

$$y = \ldots (x - \ldots) + \ldots \qquad\qquad (-0.46)(x - 15.58) + 6.08$$

i.e. $\qquad\qquad y = \ldots . \qquad\qquad 13.25 - 0.46x$

Note
As in the calculations of means and standard deviations the choice of transformation is purely arbitrary.

19

Calculate the regression line for the data given below.

The following table gives the marks x obtained by the students at an examination in arithmetic at the end of one term together with the mark y obtained at the end of the following term.

student	1	2	3	4	5	6	7	8	9	10	11	12
x	53	74	48	71	66	60	47	72	48	65	80	40
y	41	50	44	38	41	48	45	57	36	46	50	47

Find the regression line for y on x.

A complete solution is given below.

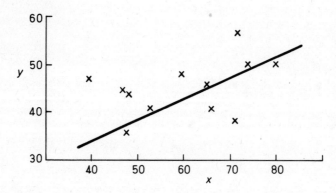

The line slopes forwards.
Therefore the value of b should be positive.

x	y	$u = x - 60$	$v = y - 45$	u^2	uv
53	41	−7	−4	49	28
74	50	14	5	196	70
48	44	−12	−1	144	12
71	38	11	−7	121	−77
66	41	6	−4	36	−24
60	48	0	3	0	0
47	45	−13	0	169	0
72	57	12	12	144	144
48	36	−12	−9	144	108
65	46	5	1	25	5
80	50	20	5	400	100
40	47	−20	2	400	−40
		$\Sigma u = 68 - 64$ $= 4$	$\Sigma v = 28 - 25$ $= 3$	$\Sigma u^2 = 1828$	$\Sigma uv = 467 - 141$ $= 326$

$$b = \frac{326 - (4 \times 3/12)}{1828 - (4 \times 4/12)} = \frac{325}{1828 - 1.33} = \frac{325}{1827}$$

$$= \underline{0.18} \text{ (to 2 d.pl.)}$$

which agrees with the fact that the line slopes upwards.

$$\bar{u} = \frac{4}{12} \qquad \bar{v} = \frac{3}{12}$$
$$= 0.33 \qquad = 0.25$$

Therefore,

$$\bar{x} = 0.33 + 60 \qquad \bar{y} = 0.25 + 45$$
$$\bar{x} = 60.33 \qquad \bar{y} = 45.25.$$

Therefore regression of y on x is

$$y = 0.18(x - 60.33) + 45.25$$
$$\underline{y = 34.39 + 0.18x}.$$

Regression coefficient

As we have already stated, the equation of a straight line is of the form

$$y = A + Bx$$

and the sign of the constant A gives the direction of slope.
We must now introduce a further fact.

The numerical value of the constant B gives the degree of slope of the straight line.

The following diagrams show straight lines and their corresponding values.

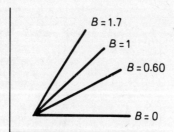 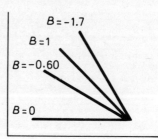

1. The equation of a straight line is of the form $y = ...$ and the slope is given by the constant	$y = A + Bx$ A
2. Similarly, the regression of x on y is given by $y = ...$ and so the slope of the regression line is given by the constant ($\bar{y}/b/b\bar{x}$).	$\bar{y} + b(x - \bar{x})$ b
3. **The constant b is known as the REGRESSION COEFFICIENT.** For example in the problem concerning the effect of temperature on the yield of a chemical process (p. 22) the regression of yield on temperature was found to be $$y = 1.44x + 9.27$$ Therefore the regression coefficient is	 1.44
4. The regression coefficient gives the ... of the regression line (see Frame 2 if in difficulty). **i.e. it gives the increase in y for a unit increase in x.**	slope
5. Consider the example given in Frame 3. The regression coefficient is 1.44 which indicates a unit increase in (yield/temperature) will be accompanied by an increase in (yield/temperature) of 1.44.	 temperature yield

6.

Give the regression coefficients for the following examples and also the interpretation you would place on the coefficients.

1. The regression of number of pupils passing 3 GCE subjects on number of pupils per teacher is

$$y = 13.25 - 0.46x$$

Regression coefficient is

0.46

Indicates that increase of 1 pupil per teacher is accompanied by a <u>decrease</u> of 0.46 in the number of pupils passing.

2. The regression of the evaporation loss from drums of aviation spirit on the length of time the drums have been stored is

$$y = 0.88 + 2.14x.$$

Regression coefficient is

2.14

Indicates that unit increase in time is accompanied by an increase in evaporation loss of 2.14.

Summary

To calculate regression of y on x.

1. (i) Plot a scatter diagram; from this it can be ascertained whether a relationship between the two variables does in fact exist.
   (ii) Draw in 'rough' regression line; if the line slopes 'upwards' the calculated value of b should be positive. If the line slopes 'downwards' the calculated value of b should be negative.

2. Calculate $b, \bar{y}, \bar{x}$

where $b = \dfrac{\Sigma xy - \dfrac{\Sigma x \Sigma y}{n}}{\Sigma x^2 - \dfrac{(\Sigma x)^2}{n}}$.

The regression of y on x is then given by $y = b(x - \bar{x}) + \bar{y}$.

3. If it appears that the calculations of $b, \bar{y}$ and $\bar{x}$ will be laborious, the variables x and y can be transformed to new variables u and v (respectively) by subtracting a constant

i.e. $u = x - A$

$v = y - B.$

The equation for b then becomes

$$b = \frac{\Sigma uv - \dfrac{\Sigma u \Sigma v}{n}}{\Sigma u^2 - \dfrac{(\Sigma u)^2}{n}}$$

and $\bar{x} = \bar{u} + A$, $\bar{y} = \bar{v} + B$.

4. (i) The constant b is known as the COEFFICIENT OF REGRESSION.
   (ii) This coefficient gives the rate of increase in y for a unit increase x.

Coefficient of correlation

Introduction

By plotting a scatter diagram we can see at a glance whether or not a relationship exists between two variables.

However we may be required to state whether or not a <u>significant</u> relationship exists between the two variables. This we cannot do merely by inspection; we, in fact, require some measure of the association between the variables.

Such a measure is the COEFFICIENT OF (LINEAR) CORRELATION which is denoted by the symbol r and calculated using the formula

$$r = \frac{\Sigma xy - \dfrac{\Sigma x \Sigma y}{n}}{\sqrt{\left[\left(\Sigma x^2 - \dfrac{(\Sigma x)^2}{n}\right)\left(\Sigma y^2 - \dfrac{(\Sigma y)^2}{n}\right)\right]}}.$$

Fig. 2.1 on p. 33 shows four scatter diagrams together with the corresponding r values.

It must be emphasised that r is a measure of a <u>linear</u> relationship

For example, consider the scatter diagram given in Fig. 2.2. This data produced a value of r equal to zero even though it is apparent that a relationship does exist between age and strength of grip. Therefore it must always be remembered that r is not an appropriate measure for non-linear relationships.

It is apparent that the evaluation of r is similar to that of b in the regression equation. In fact if the regression of y on x has already been derived then it is relatively easy to calculate r as the values of Σxy, Σx, Σy and Σx^2 will have been calculated and so we will merely have to calculate Σy^2.

Having calculated the value of r we then use the table on page 33 to test for significance. If the calculated value of r lies outside the tabulated values, we say that the result is significant at the 5 per cent level.

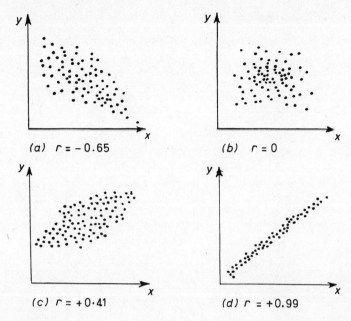

Fig. 2.1

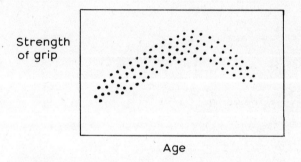

Fig. 2.2

	Significant values of r	
n	r	
10	−0.632	+0.632
20	−0.444	+0.444
30	−0.360	+0.360
40	−0.312	+0.312
50	−0.278	+0.278
60	−0.254	+0.254
70	−0.235	+0.235
80	−0.219	+0.219
90	−0.208	+0.208
100	−0.197	+0.197

Some calculations

I. Firstly we will consider the example concerning the effect of temperature on the yield of a chemical process.

This example was used earlier in the section and the following table and scatter diagram were constructed.

Table 2.3

x	y	x^2	xy	y^2	
−5	1	25	−5		1
−4	5	16	−20		25
−3	4	9	−12		16
−2	7	4	−14		49
−1	10	1	−10		100
0	8	0	0		64
1	9	1	9		81
2	13	4	26		169
3	14	9	42		196
4	13	16	52		169
5	18	25	90		324
$\Sigma x = 0$	$\Sigma y = 102$	$\Sigma x^2 = 110$	$\Sigma xy = 219$ -61 158	$\Sigma y^2 = ...$	1194

1.

The formula for the coefficient of correlation is

$$... = \frac{\left(\Sigma xy - \dfrac{\Sigma x \Sigma y}{n}\right)}{\sqrt{\left[\left(\Sigma x^2 - \dfrac{(\Sigma x)^2}{n}\right)\left(\Sigma y^2 - \dfrac{(\Sigma y)^2}{n}\right)\right]}}$$

r

In finding the regression of y on x we have already calculated values of $\Sigma xy, \Sigma x, \Sigma y, \Sigma x^2$.
Therefore we only have to find ...
Complete column 5, Table 2.3.

Σy^2

2.

Substituting numerical values into the formula for r gives

$$r = \frac{(... - ...)}{\sqrt{(... - ...)(... - ...)}}$$

$$\frac{158 - 0 \times \dfrac{102}{11}}{\sqrt{\left[\left(110 - \dfrac{(0)^2}{11}\right)\left(1194 - \dfrac{(102)^2}{11}\right)\right]}}$$

which gives

$$r = 0.96.$$

3.

We now have to interpret this result. Referring to the table of significant values of r, p. 33, we see that when $n = 10$, the significant values of r are ... and ...	-0.632 and $+0.632$
Therefore the significant values for $n = 11$ (will/will not) be approximately equal to these values.	will
The value $r = 0.96$ (does/does not) lie outside the limits.	does
Therefore we can say there (is/is not) a significant linear relationship between temperature and the yield.	is

II. Compute and interpret the correlation coefficient for the following grades of 6 students selected at random

Maths grade (x)	70	92	80	74	65	83
English grade (y)	74	84	63	87	78	90

4.

Firstly plot the scatter diagram

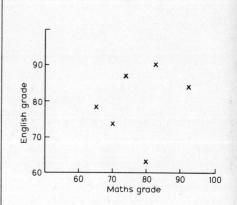

5.

From this diagram it would appear that there (is a/is no) relationship between the two sets of grades.	is no

6.

In order to confirm or reject this we calculate the	coefficient of correlation

7.

It is apparent that calculation of the correlation coefficient for this data will be very laborious; we must therefore transform the data
We will use the transformations
$$u = x - 75$$
$$v = y - 74.$$

8.

The formula for the coefficient of correlation is

$$r = \frac{\left(\Sigma xy - \dfrac{\Sigma x \Sigma y}{n}\right)}{\sqrt{\left(\Sigma x^2 - \dfrac{(\Sigma x)^2}{n}\right)\left(\Sigma y^2 - \dfrac{(\Sigma y)^2}{n}\right)}}$$

When using transformed data we merely replace Σx by Σu, Σy by Σv, Σxy by Σuv etc.
Hence the formula becomes

$$r = \frac{\left(\Sigma uv - \dfrac{\Sigma u \Sigma v}{n}\right)}{\sqrt{\left(\Sigma u^2 - \dfrac{(\Sigma u)^2}{n}\right)\left(\Sigma v^2 - \dfrac{(\Sigma v)^2}{n}\right)}}$$

9.

Complete Table 2.4, below.

x	$u = x - 75$	u^2	y	$v = y - 74$	v^2	uv
70	−5	25	74	0	0	0
92	17	289	84	10	100	170
80			63			
74			87			
65			78			
83	8	64	90	16	256	128
	$\Sigma u = ...$	$\Sigma u^2 = ...$		$\Sigma v = ...$	$\Sigma v^2 = ...$	$\Sigma uv = ...$

u	u^2	v	v^2	uv
5	25	−11	121	−55
−1	1	13	169	−13
−10	100	4	16	−40
$\Sigma u = 14$	$\Sigma u^2 = 504$	$\Sigma v = 32$	$\Sigma v^2 = 662$	$\Sigma uv = 190$

10.

The formula for r for transformed data is

$$r = \frac{\left(\Sigma uv - \dfrac{\Sigma u \Sigma v}{n}\right)}{\sqrt{\left(\Sigma u^2 - \dfrac{(\Sigma u)^2}{n}\right)\left(\Sigma v^2 - \dfrac{(\Sigma v)^2}{n}\right)}}$$

Substituting the appropriate values from Table 2.4 gives

$$r = \frac{(\ldots - \ldots)}{\sqrt{(\ldots - \ldots)(\ldots - \ldots)}}$$

$$\frac{190 - \dfrac{14 \times 32}{6}}{\sqrt{\left(504 - \dfrac{(14)^2}{6}\right)\left(662 - \dfrac{(32)^2}{6}\right)}}$$

which gives $r = 0.24$ (to 2 d.pl.).

11.

The significant values of r for $n = 6$ (are/are not) tabulated. However the significant values for $n = 10$ are $r = ...$ and $r = ...$ and from the table we see that as n (increases/decreases) the values of r increase in magnitude. Hence, for $n = 6$, the value of $r = 0.2$ (will/will not) be significant.	are not $r = -0.632$ and $r = +0.632$ decreases will not (since it will obviously lie inside the significant values).

12.

Thus, as indicated by the scatter diagram, there (is/is no) significant relationship between the two sets of grades.	is no

13.

On the basis of the obtained data (below) an experimenter asserts that the older a child is the fewer irrelevant responses he makes in an experimental situation (i.e. that there is a significant negative correlation between age and number of irrelevant responses).
Determine whether this conclusion is valid.
Give full working.
A complete solution is given below.

Age	No. of irrelevant responses	Age	No. of irrelevant responses
2	11	7	7
3	12	7	12
4	10	9	8
4	13	9	7
5	11	10	3
5	9	11	6
6	10	11	5
		12	5

Solution to exercise given in Frame 13.

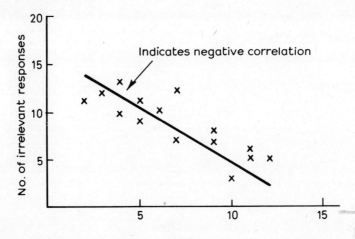

x	u = x − 6	u²	y	v = y − 8	v²	uv
2	−4	16	11	3	9	−12
3	−3	9	12	4	16	−12
4	−2	4	10	2	4	−4
4	−2	4	13	5	25	−10
5	−1	1	11	3	9	−3
5	−1	1	9	1	1	−1
6	0	0	10	2	4	0
7	1	1	7	−1	1	−1
7	1	1	12	4	16	4
9	3	9	8	0	0	0
9	3	9	7	−1	1	−3
10	4	16	3	−5	25	−20
11	5	25	6	−2	4	−10
11	5	25	5	−3	9	−15
12	6	36	5	−3	9	−18
	Σu = 28 − 13 15	Σu² = 157		Σv = 24 − 15 9	Σv² = 133	Σuv = 4 −109 −105

$$r = \frac{\left(\Sigma uv - \frac{\Sigma u \Sigma v}{n}\right)}{\sqrt{\left(\Sigma u^2 - \frac{(\Sigma u)^2}{n}\right)\left(\Sigma v^2 - \frac{(\Sigma v)^2}{n}\right)}}$$

$$r = \frac{-105 - \frac{15 \times 9}{15}}{\sqrt{\left(157 - \frac{(15)^2}{15}\right)\left(133 - \frac{(9)^2}{15}\right)}} = \frac{-114}{\sqrt{142 \times 127.6}}$$

$$= -0.85 \text{ (to 2 d.pl.)}.$$

The significant values of r for n = 10, are −0.632 and +0.632.
The value r = −0.85 lies outside these limits.
Therefore it must also lie outside the limits for n = 15 as these will be less in magnitude than 0.632.
Hence there is a significant relationship between age and the number of irrelevant responses.

Consider the following three sets of data.

14.

The raw data for the experiment given in Frame 13 is shown on p. 39.

Result	Age	No. of irrelevant
1	2	11
2	7	12
3	4	13
4	10	3
5	12	5
6	3	12
7	6	10
8	7	7
9	9	7
10	11	6
11	4	10
12	5	11
13	5	9
14	9	8
15	11	5

We will consider the following three sets of data.
(i) results 1 – 5
(ii) results 1 – 10
(iii) results 1 – 15.

15.

In the diagram below we have plotted the scatter diagrams for the three sets of data.

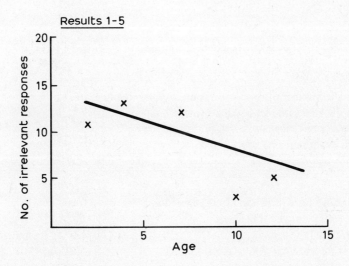

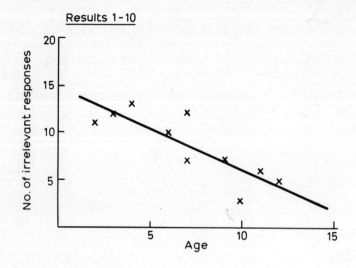

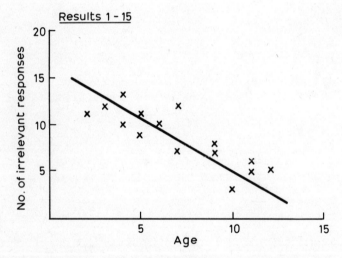

It can be seen that as the number of observations increases so the relationship between the two variables becomes (more/less) apparent, i.e. we would expect the coefficient of correlation to (vary/remain constant) as n increases.	more vary

16.

On the scatter diagrams we have drawn in *rough* regression lines. It can be seen that the slopes of the three lines (are approximately/are not) the same. Thus we would expect the coefficient of regression b to (vary/remain approximately constant) as n increases.	are approximately remain approximately constant

17.

The following table of b and r for $n = 5, 10, 15$ confirms the above statements.

n	b	r	
5	−0.88	−0.81	significance doubtful.
10	−0.84	−0.84	significant at 5% level.
15	−0.80	−0.85	significant at 5% level.

Summary

1. The coefficient of correlation is a measure of the association between two variables and using this we can determine whether a *significant linear* relationship exists between the two variables.

2. The coefficient of correlation is denoted by the symbol, r and is calculated using the formula

$$r = \frac{\Sigma xy - \dfrac{\Sigma x \Sigma y}{n}}{\sqrt{\left[\left(\Sigma x^2 - \dfrac{(\Sigma x)^2}{n}\right)\left(\Sigma y^2 - \dfrac{(\Sigma y)^2}{n}\right)\right]}}$$

3. Having calculated the value of r, we use the table of significant value of r, (given on p. 33) to determine whether the calculated value is significant — a value of r outside the given range indicates a significant linear relationship.

   NOTE: r is only a measure of a *LINEAR* relationship.

4. If transformed variables $u = x - A$ are used the equation for r becomes
$v = y - B$

$$r = \frac{\Sigma uv - \dfrac{\Sigma u \Sigma v}{n}}{\sqrt{\left[\left(\Sigma u^2 - \dfrac{(\Sigma u)^2}{n}\right)\left(\Sigma v^2 - \dfrac{(\Sigma y)^2}{n}\right)\right]}}$$

5. As n increases, the coefficient of correlation varies — it becomes more significant — whereas the coefficient of regression remains approximately constant.

Exercises for Section 2

PHYSICAL SCIENCES AND ENGINEERING

Regression

1. The height of soapsuds in the sink is of importance to soap manufacturers. An experiment was performed by varying the amount of soap and measuring the height of the suds in a standard dishpan after a given amount of agitation. The data are as follows:-

gm of product	4.0	4.5	5.0	5.5	6.0	6.5	7.0
suds height	33	42	45	51	53	61	62

   Calculate the regression of suds height on gm of product.

2. The carbon content in a clay was estimated by different methods. Method A was direct and an accurate measurement; method B was indirect and an approximation. The results are as follows:-

Method A	1.2	1.5	1.4	2.4	1.9	2.0	2.6	2.9	1.6	1.3
Method B	0.9	1.0	1.5	1.8	1.4	1.3	2.6	2.8	1.1	0.9

   Calculate the regression of the results obtained by method B on those obtained by method A.
   State the estimated regression coefficient. What interpretation would you place on this coefficient?
   Does this coefficient have the numerical value you would expect? Give reasons.

Correlation

1. A student obtained the following data concerning the amount of potassium bromide, y, which will dissolve in 100 gm of water at various temperatures, x.

x (°C)	0	10	20	30	40	50
y (gm)	52	60	64	73	76	81

   Calculate the coefficient of correlation, r, and carry out the appropriate test of significance of r.

2. Intelligence test scores and freshmen Chemistry grades.

Student	Test score (x)	Chem. grade (y)
1	65	85
2	50	74
3	55	76
4	65	90
5	55	85
6	70	87
7	65	94
8	70	98
9	55	81
10	70	91
11	50	76
12	55	74

   Is there a significant relationship between the two grades?

BIOLOGICAL SCIENCES

Regression

1. The height of soapsuds in the sink is of importance to soap manufacturers. An experiment was performed by varying the amount of soap and measuring the height of the suds in a standard dishpan after a given amount of agitation. The data are as follows:-

gm of product	4.0	4.5	5.0	5.5	6.0	6.5	7.0
suds height	33	42	45	51	53	61	62

   Calculate the regression of suds height on gm of product.

2. 10 normal rabbits were selected at random and not subjected to any treatment. The anterior muscles of both hind legs of each rabbit were removed and weighed. The results were as follows:-

Rabbit No.	Muscle weights (gm) Anterior	
	Left leg x	Right leg y
1	5.0	4.9
2	4.8	5.0
3	4.3	4.3
4	5.1	5.3
5	4.1	4.1
6	4.0	4.0
7	7.1	6.9
8	5.9	6.3
9	5.3	5.2
10	5.3	5.5

Calculate the regression of y on x.

State the estimated regression coefficient. What interpretation would you place on this coefficient? Does this coefficient have the numerical value you would expect? Give resons.

Coefficient of correlation

1. The table gives the weight of heart (x), and the weight of kidneys (y), in a random sample of 12 adult males between the ages of 25 and 55 years.

Male No.	Heart weight (in oz) x	Kidney weight (in oz) y
1	11.50	11.25
2	9.50	11.75
3	13.00	11.75
4	15.50	12.50
5	12.50	12.50
6	11.50	12.75
7	9.00	9.50
8	11.50	10.75
9	9.25	11.00
10	9.75	9.50
11	14.25	13.00
12	10.75	12.00

$\Sigma x = 138.00$ $\quad \Sigma y = 138.25$

$\Sigma xy = 1608.12$ $\quad \Sigma x^2 = 1632.75$ $\quad \Sigma y^2 = 1607.81$

Calculate the coefficient of correlation, r, and carry out the appropriate test of significance on r.

2. The table below gives the ages, x, and the blood pressure, y, of a sample of eight women.

Age, x.	43	36	63	49	60	38	72	55
Blood pressure, y.	140	116	150	143	155	116	158	150

Is there a significant relationship between age and blood pressure?

SOCIAL SCIENCES

Regression

1. The height of soap suds in the sink is of importance to soap manufacturers. An experiment was performed by varying the amount of soap and measuring the height of the soap suds in a standard dishpan after a given amount of agitation. The data are as follows:

gm of product	4.0	4.5	5.0	5.5	6.0	6.5	7.0
suds height	33	42	45	51	53	61	62

Calculate the regression of suds height on gm of product.

2. A psychological study involved the rating of rats along a dominance-submissiveness continuum. In order to determine the reliability of the rating, the scores given by two different observers were tabulated. Calculate the regression of the scoring of observer B on that of observer A.

	Rat no.	1	2	3	4	5	6	7	8	9	10
Score	Observer A	12	2	3	1	4	5	14	11	9	7
	Observer B	13	1	5	2	2	3	11	10	9	6

State the estimated regression coefficient. What interpretation would you place on this coefficient? Does this coefficient have the numerical value (approx.) you would expect?

Give reasons.

Coefficient of Correlation

1. In trying to evaluate the effectiveness of its advertising a firm computed the following information.

Year	x Advertising expenditure	y Revenue (£00 000's)
1958	6	2.5
1959	12	3.8
1960	15	3.9
1961	15	4.2
1962	23	5.0
1963	24	4.8
1964	38	6.2
1965	42	6.0
1966	47	7.9

$\Sigma xy = 1271.21 \qquad \Sigma x = 222 \qquad \Sigma y = 44.30$
$\qquad\qquad\qquad\quad \Sigma x^2 = 7152 \qquad \Sigma y^2 = 238.43$

Calculate the coefficient of correlation, r, and carry out the appropriate test of significance on r.

2. Calculate the coefficient of correlation between the following series of male and female mortality rates per 10 000 of population.

Year	Male rate	Female rate
1935	125	111
1936	129	114
1937	132	117
1938	125	108
1939	130	113
1940	161	129
1941	157	118
1942	144	107
1943	153	113
1944	153	108

Is there a significant relationship between the two rates?

Solutions

Section 1

PHYSICAL SCIENCES AND ENGINEERING

1. Null hypothesis: Assume $\sigma = 0.01$

    Calculate χ^2 using $\chi^2 = \dfrac{(n-1)s^2}{\sigma^2}$

    where $s^2 = 0.0004$, $\sigma^2 = (0.01)^2$, $n = 12$

    $$\chi^2 = \frac{11 \times 0.0004}{0.0001}$$

    $$= 11 \times 4$$

    $$\chi^2 = 44.$$

    Test using χ^2 distribution with 11 d.f.

    1% significance level $\chi^2 = 2.603$ and $\chi^2 = 26.757$
    5% significance level $\chi^2 = 3.816$ and $\chi^2 = 21.920$.

    The value $\chi^2 = 44$ is significant at 1 per cent level.
    Almost conclusive evidence that $\sigma \neq 0.01$ and we reject the null hypothesis that $\sigma = 0.01$.

2. Null hypothesis: Assume $\sigma_1 = \sigma_2$

    Calculate F using $F = \dfrac{\text{larger sample variance}}{\text{smaller sample variance}}$.

    In this example $s_1 = 10$, $s_2 = 18$, $n_1 = 25$, $n_2 = 20$.

    Therefore $F = \dfrac{(18)^2}{(10)^2} = 3.24$.

Test using F distribution with $(20-1)$ and $(25-1)$ d.f.

    1% significance level $F \simeq 2.74$
    5% significance level $F \simeq 2.03$.

The value $F = 3.24$ is significant at 1 per cent level.
Therefore there is almost conclusive evidence of a real difference between σ_1 and σ_2 and we reject the null hypothesis.

3. Null hypothesis: $\sigma^2 = 0.16$
   Alternative hypothesis: $\sigma^2 < 0.16$

Calculate χ^2 using $\chi^2 = \dfrac{(n-1)s^2}{\sigma^2}$

where $s^2 = 0.11, \; \sigma^2 = 0.16, \; n = 25$

Therefore $\chi^2 = \dfrac{24 \times 0.11}{0.16} = 16.5,$

Test using χ^2 distribution with 24 d.f.

    1% significance level $\chi^2 = 10.856$
    5% significance level $\chi^2 = 13.848$.

The value $\chi^2 = 16.5$ is not significant at the 5 per cent level.
Therefore there is no evidence that $\sigma^2 < 0.16$ and we accept the null hypothesis.
i.e. we assume that the inspector is making satisfactory measurements.

4. Null hypothesis: Assume $\sigma_A^2 = \sigma_B^2$

Calculate F using $F = \dfrac{\text{larger sample variance}}{\text{smaller sample variance}}$

where $s_A^2 = 18.7$ $s_B^2 = 4.5$
              $n_A = 6$ $n_B = 8$

Therefore $F = \dfrac{18.7}{4.5}$

               $= 4.16.$

Test using F distribution with 5 and 7 d.f.

    1% significance level $F = 7.46$
    5% significance level $F = 3.97$.

The value F is significant at 5 per cent level.
Therefore reasonable evidence that $\sigma_A^2 \neq \sigma_B^2$ and we reject the null hypothesis.
i.e. we presume that B is the more precise observer.

BIOLOGICAL SCIENCES

1. See physical sciences and engineering; no: 1. (p. 47)

2. See physical sciences and engineering; no: 2. (p. 47)

3. Null hypothesis: $\sigma = 4$
Alternative hypothesis: $\sigma > 4$

Calculate χ^2 using $\chi^2 = \dfrac{(n-1)s^2}{\sigma^2}$

In this example $n = 20$, $s^2 = (5)^2$, $\sigma^2 = (4)^2$

Therefore $\chi^2 = \dfrac{19 \times 25}{16} = 29.7$

Test using χ^2 distribution with 19 d.f. (one-tailed test)

1% significance level $\chi^2 = 36.191$
5% significance level $\chi^2 = 30.144$

The value $\chi^2 = 29.7$ is not significant at 5 per cent level and we accept the null hypothesis (with some reservations)
Therefore we conclude that there is reasonable evidence that the standard deviation of increase in pulse rate for teenagers is 4.

4. Null hypothesis: Assume $\sigma_1^2 = \sigma_2^2$

Calculate F using $F = \dfrac{\text{larger sample variance}}{\text{smaller sample variance}}$.

In this example $n_1 = 10$, $s_1^2 = 0.0862$
$n_2 = 10$, $s_2^2 = 0.0672$.

Therefore $F = \dfrac{0.0862}{0.0672}$

$= 1.28$.

Test using F distribution with 9 and 9 d.f.

1% significance level $F = 5.35$
5% significance level $F = 3.18$

The value $F = 1.28$ is not significant at the 5% level.
Therefore there is no evidence that $\sigma_1^2 \neq \sigma_2^2$ and we accept the null hypothesis that $\sigma_1^2 = \sigma_2^2$.

SOCIAL SCIENCES

1. See physical science and engineering; no: 1. (p. 47)

2. See physical sciences and engineering; no: 2. (p. 47)

3. Null Hypothesis : $\sigma = 1$
    Alternative Hypothesis : $\sigma < 1$

    Calculate χ^2 using $\chi^2 = \dfrac{(n-1)s^2}{\sigma^2}$ where $s^2 = (0.7)^2$,
    $$\sigma^2 = (1)^2,$$
    $$n = 20.$$

    Therefore $\chi^2 = 19 \times 0.49 = 9.31$.
    Test, using χ^2 distribution with 19 d.f.

        1% significance level $\chi^2 = 7.633$
        5% significance level $\chi^2 = 10.117$.

    The value $\chi^2 = 9.31$ is significant at 5% level
    ∴ reject null hypothesis that $\sigma = 1$.
    i.e. there is reasonable evidence to assume that for the new process $\sigma < 1$.

4. Null Hypothesis: Assume $\sigma_1^2 = \sigma_2^2$

    Calculate F where $F = \dfrac{\text{larger sample variance}}{\text{smaller sample variance}}$.

    In this example, $n_1 = 11$, $s_1 = 3$,
    $$n_2 = 16, \quad s_2 = 6$$

    Therefore $\quad F = \dfrac{(6)^2}{(3)^2} = 4$.

    Test, using F distribution with 15 and 10 d.f.

        1% level of significance $F = 4.56$
        5% level of significance $F = 2.85$.

    The value $F = 4$ is significant at 5% level.
    Therefore reasonable evidence to assume that $\sigma_1^2 \neq \sigma_2^2$ and we reject the null hypothesis.

Section 2

PHYSICAL SCIENCES AND ENGINEERING

Regression

1.

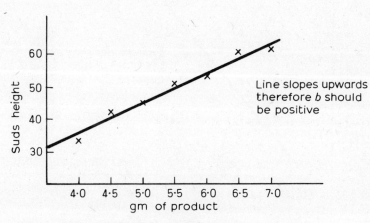

Line slopes upwards therefore b should be positive

x	y	$u = x - 5.5$	$v = y - 50$	u^2	uv
4.0	33	−1.5	−17	2.25	25.5
4.5	42	−1.0	−8	1.00	8.0
5.0	45	−0.5	−5	0.25	2.5
5.5	51	0	1	0.00	0.0
6.0	53	0.5	3	0.25	1.5
6.5	61	1.0	11	1.00	11.0
7.0	62	1.5	12	2.25	18.0
		$\Sigma u = 0$	$\Sigma v = 27$ -30 -3	$\Sigma u^2 = 7$	$\Sigma uv = 66.5$

$$b = \frac{\Sigma uv - \dfrac{\Sigma u \Sigma v}{n}}{\Sigma u^2 - \dfrac{(\Sigma u)^2}{n}} = \frac{66.5 - \dfrac{0 \times (-3)}{7}}{7 - \dfrac{(0)^2}{7}} = \frac{66.5}{7} = 9.50 \text{ (to 2 d.pl.)}$$

which agrees with the fact that the 'rough' regression line slopes 'upwards'.

$$\bar{u} = 0 \qquad \bar{v} = \frac{-3}{7} = -0.43$$

Therefore $\bar{x} = 0 + 5.5, \quad \bar{y} = -0.43 + 50$
$\qquad\qquad = 5.5 \qquad\qquad = 49.57$

Regression of y on x is given by

$$y = b(x - \bar{x}) + \bar{y}$$
$$= 9.50(x - 5.5) + 49.57$$

therefore $y = 9.50x - 2.68$.

2.

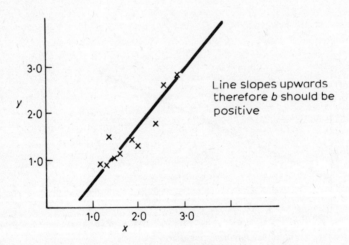

Line slopes upwards therefore b should be positive

Method A x	Method B y	$u = x - 2.0$	$v = y - 2.0$	u^2	uv
1.2	0.9	−0.8	−1.1	0.64	0.88
1.5	1.0	−0.5	−1.0	0.25	0.50
1.4	1.5	−0.6	−0.5	0.36	0.30
2.4	1.8	0.4	−0.2	0.16	−0.08
1.9	1.4	−0.1	−0.6	0.01	0.06
2.0	1.3	0	−0.7	0.00	0.00
2.6	2.6	0.6	0.6	0.36	0.36
2.9	2.8	0.9	0.8	0.81	0.72
1.6	1.1	−0.4	−0.9	0.16	0.36
1.3	0.9	−0.7	−1.1	0.49	0.77
		$\Sigma u = 1.9$ $\underline{-3.1}$ $= -1.2$	$\Sigma v = 1.4$ $\underline{-6.1}$ $= -4.7$	$\Sigma u^2 = 3.24$	$\Sigma uv = 3.95 - 0.08$ $= 3.87$

$$b = \frac{\Sigma uv - \dfrac{\Sigma u \Sigma v}{n}}{\Sigma u^2 - \dfrac{(\Sigma u)^2}{n}} = \frac{3.87 - \dfrac{(-1.2)(-4.7)}{10}}{3.24 - \dfrac{(-1.2)^2}{10}} = \frac{3.306}{3.096}$$

$= 1.068$ which agrees with the fact that the 'rough' regression line slopes 'upwards'.

$$\bar{u} = -0.12 \qquad \bar{v} = -0.47$$
$$\bar{x} = -0.12 + 2.0 \qquad \bar{y} = -0.47 + 2.0$$
$$= 1.88 \qquad = 1.53.$$

Regression of y on x is given by $y = b(x - \bar{x}) + \bar{y}$

$$y = 1.07(x - 1.88) + 1.53$$
$$= 1.07x + 0.48.$$

Regression coefficient is $b = 1.07$
i.e. a unit increase in carbon content as obtained by method A is accompanied by an increase of 1.07 in carbon content as obtained by method B.
Yes, the coefficient does have the expected numerical value (approx.).

In theory the regression coefficient should equal 1. Since for unit increase in carbon content as obtained by method A, the expected increase in that as obtained by method B is 1.

Correlation

1.

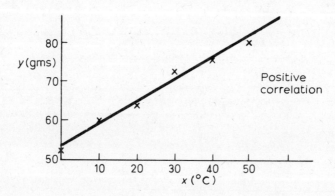

x	y	$u = x - 30$	$v = y - 70$	u^2	v^2	uv
0	52	−30	−18	900	324	540
10	60	−20	−10	400	100	200
20	64	−10	−6	100	36	60
30	73	0	3	0	9	0
40	76	10	6	100	36	60
50	81	20	11	400	121	220
		$\Sigma u = -30$	$\Sigma v = -14$	$\Sigma u^2 = 1900$	$\Sigma v^2 = 626$	$\Sigma uv = 1080$

$$r = \frac{1080 - \frac{30 \times 14}{6}}{\sqrt{\left[\left(1900 - \frac{(30)^2}{6}\right)\left(626 - \frac{(14)^2}{6}\right)\right]}}$$

$$= \frac{1010}{\sqrt{1750 \times 593.33}}.$$

$$\underline{r = 0.99} \text{ (to 2 d.pl.).}$$

The significant values of r for $n = 10$ are -0.632 and $+0.632$.
Therefore it would appear that for $n = 6$ the value $r = 0.99$ is significant.
i.e. there is a significant relationship between the amount of potassium bromide which will dissolve in 100 gm of water and the temperature of the water.

2.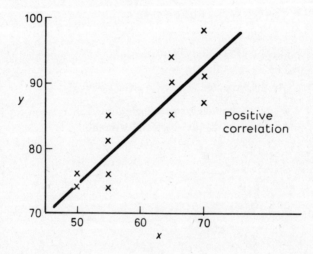

Positive correlation

x	y	$u = x - 60$	$v = y - 85$	u^2	v^2	uv
65	85	5	0	25	0	0
50	74	−10	−11	100	121	110
55	76	−5	−9	25	81	45
65	90	5	5	25	25	25
55	85	−5	0	25	0	0
70	87	10	2	100	4	20
65	94	5	9	25	81	45
70	98	10	13	100	169	130
55	81	−5	−4	25	16	20
70	91	10	6	100	36	60
50	76	−10	−9	100	81	90
55	74	−5	−11	25	121	55
		$\Sigma u = 5$	$\Sigma v = 35$ −44 − 9	$\Sigma u^2 = 675$	$\Sigma v^2 = 735$	$\Sigma uv = 600$

$$r = \frac{600 - \frac{(5)(-9)}{12}}{\sqrt{\left[\left(675 - \frac{(5)^2}{12}\right)\left(735 - \frac{(-9)^2}{12}\right)\right]}}$$

$$= \frac{603.75}{\sqrt{672.92 \times 728.25}}$$

$$= 0.86 \text{ (to 2 d.pl.)}.$$

The significant values of r for $n = 10$ are -0.632 and $+0.632$.
Therefore for $n = 12$ the value $r = 0.86$ is significant

i.e. there is a significant relationship between the two grades.

BIOLOGICAL SCIENCES

Regression

1. See physical sciences and engineering — Regression; no: 1.

2.

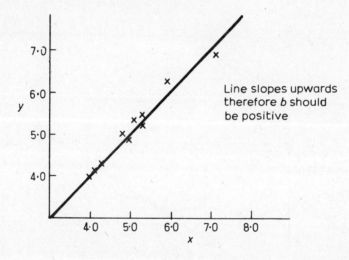

Line slopes upwards therefore b should be positive

x	y	$u = x - 5.0$	$v = y - 5.0$	u^2	uv
5.0	4.9	0	−0.1	0	0
4.8	5.0	−0.2	0	0.04	0
4.3	4.3	−0.7	−0.7	0.49	0.49
5.1	5.3	0.1	0.3	0.01	0.03
4.1	4.1	−0.9	−0.9	0.81	0.81
4.0	4.0	−1.0	−1.0	1.00	1.00
7.1	6.9	2.1	1.9	4.41	3.99
5.9	6.3	0.9	1.3	0.81	1.17
5.3	5.2	0.3	0.2	0.09	0.06
5.3	5.5	0.3	0.5	0.09	0.15
		$\Sigma u = 3.7$	$\Sigma v = 4.2$	$\Sigma u^2 = 7.75$	$\Sigma uv = 7.70$
		-2.8	-2.7		
		$= 0.9$	$= 1.5$		

$$b = \frac{\Sigma uv - \dfrac{\Sigma u \Sigma v}{n}}{\Sigma u^2 - \dfrac{(\Sigma u)^2}{n}} = \frac{7.70 - \dfrac{0.9 \times 1.5}{10}}{7.75 - \dfrac{(0.9)^2}{10}} = \frac{7.565}{7.669} = 0.9865$$

which agrees with the fact that the 'rough' regression line slopes 'upwards'.

$$\bar{u} = 0.09 \quad \bar{v} = 0.15$$

therefore $\bar{x} = 5.09 \quad \bar{y} = 5.15.$

Regression of y on x is given by

$$y = b(x - \bar{x}) + \bar{y}$$
$$= 0.99(x - 5.09) + 5.15$$

therefore $y = 0.99x + 0.11.$

Regression coefficient is $b = 0.99$
i.e. a unit increase in the weight of left leg is accompanied by an increase of 0.99 gm in the weight of right leg. Yes (approximately).
In theory the regression coefficient should equal 1, since for a unit increase in weight of left leg, the expected increase in weight of right leg is 1 gm.

Coefficient of correlation

1.

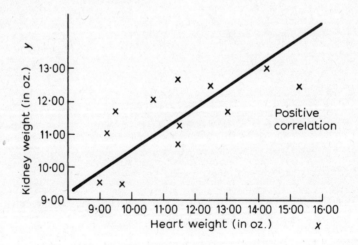

$$r = \frac{\left(\Sigma xy - \dfrac{\Sigma x \Sigma y}{n}\right)}{\sqrt{\left(\Sigma x^2 - \dfrac{(\Sigma x)^2}{n}\right)\left(\Sigma y^2 - \dfrac{(\Sigma y)^2}{n}\right)}}.$$

In this example, $\Sigma x = 138.00$, $\Sigma y = 138.25$
$\Sigma xy = 1608.12$, $\Sigma x^2 = 1632.75$, $\Sigma y^2 = 1607.81$

$$\therefore \quad r = \frac{1608.12 - \dfrac{138.00 \times 138.25}{12}}{\sqrt{\left(1632.75 - \dfrac{(138.00)^2}{12}\right)\left(1607.81 - \dfrac{(138.25)^2}{12}\right)}}$$

$$r = 0.70 \text{ (to 2 d.pl.).}$$

Significant values of r for $n = 10$ are -0.632 and $+0.632$
and for $n = 20$ are -0.444 and $+0.444$.

∴ when $n = 12$, the value $r = 0.70$ is obviously significant.
i.e. there is a significant relationship between heart weight and kidney weight.

2.

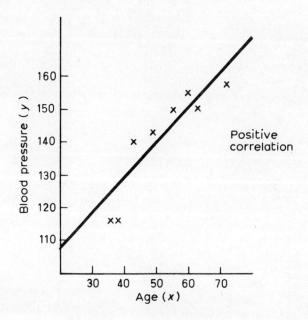

Positive correlation

Age (x)	Blood pressure (y)	$u = x - 50$	$v = y - 150$	uv	u^2	v^2
43	140	-7	-10	70	49	100
36	116	-14	-34	476	196	1156
63	150	13	0	0	169	0
49	143	-1	-7	7	1	49
60	155	10	10	50	100	25
38	116	-12	-34	408	144	1156
72	158	22	8	176	484	64
55	150	5	0	0	25	0
		$\Sigma u = 50$ -34 $\overline{16}$	$\Sigma v = 13$ -85 $\overline{-72}$	$\Sigma uv = 1187$	$\Sigma u^2 = 1168$	$\Sigma v^2 = 2550$

$$r = \frac{\left(\Sigma uv - \dfrac{\Sigma u \Sigma v}{n}\right)}{\sqrt{\left(\Sigma u^2 - \dfrac{(\Sigma u)^2}{n}\right)\left(\Sigma v^2 - \dfrac{(\Sigma v)^2}{n}\right)}}$$

$$= \frac{1187 - \dfrac{16 \times (-72)}{8}}{\sqrt{\left(1168 - \dfrac{(16)^2}{8}\right)\left(2550 - \dfrac{(72)^2}{8}\right)}}$$

$$= \frac{1131}{\sqrt{1136 \times 1902}}$$

$$= 0.91 \text{ (to 2 d.pl.).}$$

For $n = 10$ significant values of r are -0.632 and $+0.632$

∴ It appears that for $n = 8$ the value of $r = 0.91$ is significant

i.e. there is a significant relationship between age and blood pressure.

SOCIAL SCIENCES

Regression

1. See physical sciences and engineering — Regression; no: 1.

2.

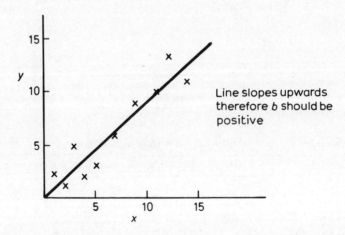

Line slopes upwards therefore b should be positive

	Rank		
Obs. A (x)	Obs. B (y)	x^2	xy
12	13	144	156
2	1	4	2
3	5	9	15
1	2	1	2
4	2	16	8
5	3	25	15
14	11	196	154
11	10	121	110
9	9	81	81
7	6	49	42
$\Sigma x = 68$	$\Sigma y = 62$	$\Sigma x^2 = 646$	$\Sigma xy = 585$

$$b = \frac{\Sigma xy - \frac{\Sigma x \Sigma y}{n}}{\Sigma x^2 - \frac{(\Sigma x)^2}{n}} = \frac{585 - \frac{68 \times 62}{10}}{646 - \frac{(68)^2}{10}} = \frac{163.4}{183.6} = 0.89$$

which agrees with the fact that the 'rough' regression line slopes upwards.

$\bar{x} = 6.8 \qquad \bar{y} = 6.2$

Regression of y on x is given by

$y = b(x - \bar{x}) + \bar{y}$

$ = 0.89(x - 6.8) + 6.2$

therefore $\quad y = 0.89x + 0.15$.

Regression coefficient is $b = 0.89$
i.e. a unit increase in rank by Observer A is accompanied by an increase of 0.89 in rank by Observer B. Yes, the coefficient does give the expected numerical value (approximately).
In theory the regression coefficient should equal 1. Since for a unit increase in rank by Observer A the expected increase in rank by Observer B is 1.

Coefficient of Correlation

1.

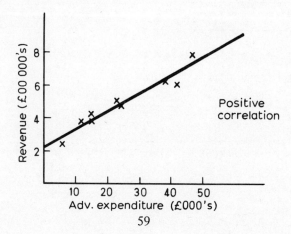

$$r = \frac{\left(\Sigma xy - \dfrac{\Sigma x \Sigma y}{n}\right)}{\sqrt{\left(\Sigma x^2 - \dfrac{(\Sigma x)^2}{n}\right)\left(\Sigma y^2 - \dfrac{(\Sigma y)^2}{n}\right)}}.$$

In this example, $\Sigma x = 222$, $\Sigma y = 44.30$

$\Sigma xy = 1271.21$, $\Sigma x^2 = 7152$, $\Sigma y^2 = 238.43$

$$\therefore \quad r = \frac{\left(1271.21 - \dfrac{222 \times 44.30}{9}\right)}{\sqrt{\left(7152 - \dfrac{(222)^2}{9}\right)\left(238.43 - \dfrac{(44.30)^2}{9}\right)}}.$$

$= 0.97$ (to 2 d.pl.).

when $n = 10$ significant values of r are -0.632 and $+0.632$
∴ it would appear that for $n = 9$ the value $r = 0.97$ is significant.
Therefore, we can say there is a significant relationship between advertising expenditure and revenue.

2.

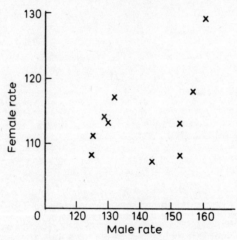

Male rate x	Female rate y	$u = x - 140$	$v = y - 110$	uv	u^2	v^2
125	111	−15	1	−15	225	1
129	114	−11	4	−44	121	16
132	117	−8	7	−56	64	49
125	108	−15	−2	30	225	4
130	113	−10	3	−30	100	9
161	129	21	19	399	441	361
157	118	17	8	136	289	64
144	107	4	−3	−12	16	9
153	113	13	3	39	169	9
153	108	13	−2	−26	169	4
		$\Sigma u = 68$	$\Sigma v = 42$	$\Sigma uv = 574$	$\Sigma u^2 = 1819$	$\Sigma v^2 = 526$
		− 59	−4	− 153		
		9	38	421		

$$\therefore \quad r = \frac{\left(421 - \dfrac{9 \times 38}{10}\right)}{\sqrt{\left(1819 - \dfrac{(9)^2}{10}\right)\left(526 - \dfrac{(38)^2}{10}\right)}}$$

$$= \frac{386.8}{\sqrt{1810.9 \times 381.6}}$$

$$= \underline{0.47} \text{ (to 2 d.pl.).}$$

When $n = 10$, significant values of r are -0.632 and $+0.632$.
$\therefore$ the value $r = 0.47$ is not significant and we can say that no significant relationship exists between male and female mortality rate.

Tables

CHI-SQUARE TABLE — TESTS INVOLVING VARIANCES

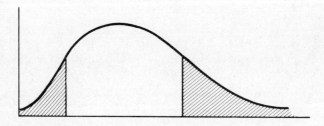

A denotes *sum* of shaded areas.

v or d.f.	$A = 0.05$		$A = 0.01$	
1	$0.0^3 982$	5.024	$0.0^4 393$	7.879
2	0.0506	7.378	0.0100	10.597
3	0.216	9.348	0.0717	12.838
4	0.484	11.143	0.207	14.860
5	0.831	12.832	0.412	16.750
6	1.237	14.449	0.676	18.548
7	1.690	16.013	0.989	20.278
8	2.180	17.535	1.344	21.955
9	2.700	19.023	1.735	23.589
10	3.247	20.483	2.156	25.188
11	3.816	21.920	2.603	26.757
12	4.404	23.337	3.074	28.300
13	5.009	24.737	3.565	29.819
14	5.629	26.119	4.075	31.319
15	6.262	27.488	4.601	32.801
16	6.908	28.845	5.142	34.267
17	7.564	30.191	5.697	35.718
18	8.231	31.526	6.265	37.156
19	8.907	32.852	6.844	38.582
20	9.591	34.170	7.434	39.997
21	10.283	35.479	8.034	41.401
22	10.982	36.781	8.643	42.796
23	11.689	38.076	9.260	44.181
24	12.401	39.364	9.886	45.558
25	13.120	40.646	10.520	46.928
26	13.844	41.923	11.160	48.290
27	14.573	43.194	11.808	49.645
28	15.308	44.461	12.461	50.993
29	16.047	45.722	13.121	52.336
30	16.791	46.979	13.787	53.672

ONE-TAILED TESTS

I.

Null hypothesis: $\sigma = \sigma_0$
Alternative hypothesis: $\sigma < \sigma_0$

	levels of significance	
d.f.	5 per cent	1 per cent
1	$0.0^2 393$	$0.0^3 157$
2	0.103	0.0201
3	0.352	0.115
4	0.711	0.297
5	1.145	0.554
6	1.635	0.872
7	2.167	1.239
8	2.733	1.646
9	3.325	2.088
10	3.940	2.558
11	4.575	3.053
12	5.226	3.571
13	5.892	4.107
14	6.571	4.660
15	7.261	5.229
16	7.962	5.812
17	8.672	6.408
18	9.390	7.015
19	10.117	7.633
20	10.851	8.260
21	11.591	8.897
22	12.338	9.542
23	13.091	10.196
24	13.848	10.856
25	14.611	11.524
26	15.379	12.198
27	16.151	12.879
28	16.928	13.565
29	17.708	14.256
30	18.493	14.953

ONE-TAILED TESTS

II.

Null hypothesis: $\sigma = \sigma_0$
Alternative hypothesis: $\sigma > \sigma_0$

d.f.	levels of significance	
	5 per cent	1 per cent
1	3.841	6.635
2	5.991	9.210
3	7.815	11.345
4	9.488	13.277
5	11.070	15.086
6	12.592	16.812
7	14.067	18.475
8	15.507	20.090
9	16.919	21.666
10	18.307	23.209
11	19.675	24.725
12	21.026	26.217
13	22.362	27.688
14	23.685	29.141
15	24.996	30.578
16	26.296	32.000
17	27.587	33.409
18	28.869	34.805
19	30.144	36.191
20	31.410	37.566
21	32.671	38.932
22	33.924	40.289
23	35.172	41.638
24	36.415	42.980
25	37.652	44.314
26	38.885	45.642
27	40.113	46.963
28	41.337	48.278
29	42.557	49.588
30	43.773	50.892

TWO-TAILED TEST

Null hypothesis: $\sigma = \sigma_0$
Alternative hypothesis: $\sigma \neq \sigma_0$

d.f.	\multicolumn{4}{c}{levels of significance}			
	\multicolumn{2}{c}{5 per cent}	\multicolumn{2}{c}{1 per cent}		
1	$0.0^3 982$	5.024	$0.0^4 393$	7.879
2	0.0506	7.378	0.0100	10.597
3	0.216	9.348	0.0717	12.838
4	0.484	11.143	0.207	14.860
5	0.831	12.832	0.412	16.750
6	1.237	14.449	0.676	18.548
7	1.690	16.013	0.989	20.278
8	2.180	17.535	1.344	21.955
9	2.700	19.023	1.735	23.589
10	3.247	20.483	2.156	25.188
11	3.816	21.920	2.603	26.757
12	4.404	23.337	3.074	28.300
13	5.009	24.736	3.565	29.819
14	5.629	26.119	4.075	31.319
15	6.262	27.488	4.601	32.801
16	6.908	28.845	5.142	34.267
17	7.564	30.191	5.697	35.718
18	8.231	31.526	6.265	37.156
19	8.907	32.852	6.844	38.582
20	9.591	34.170	7.434	39.997
21	10.283	35.479	8.034	41.401
22	10.982	36.781	8.643	42.796
23	11.689	38.076	9.260	44.181
24	12.401	39.364	9.886	45.558
25	13.120	40.646	10.520	46.928
26	13.844	41.923	11.160	48.290
27	14.573	43.194	11.808	49.645
28	15.308	44.461	12.461	50.993
29	16.047	45.722	13.121	52.336
30	16.791	46.979	13.787	53.672

F-DISTRIBUTION   1% significance level

		\multicolumn{10}{c}{Degrees of freedom for numerator}
		1	2	3	4	5	6	7	8	9	10
Degrees of freedom for denominator	1	4052	5000	5403	5625	5764	5859	5928	5982	6023	6056
	2	98.5	99.0	99.2	99.2	99.3	99.3	99.4	99.4	99.4	99.4
	3	34.1	30.8	29.5	28.7	28.2	27.9	27.7	27.5	27.3	27.2
	4	21.2	18.0	16.7	16.0	15.5	15.2	15.0	14.8	14.7	14.5
	5	16.3	13.3	12.1	11.4	11.0	10;7	10;5	10.3	10.2	10.1
	6	13.7	10.9	9.78	9.15	8.75	8.47	8.26	8.10	7.98	7.87
	7	12.2	9.55	8.45	7.85	7.46	7.19	6.99	6.84	6.72	6.62
	8	11.3	8.65	7.59	7.01	6.63	6.37	6.18	6.03	5.91	5.81
	9	10.6	8.02	6.99	6.42	6.06	5.80	5.61	5.47	5.35	5.26
	10	10.0	7.56	6.55	5.99	5.64	5.39	5.20	5.06	4.94	4.85
	11	9.65	7.21	6.22	5.67	5.32	5.07	4.89	4.74	4.63	4.54
	12	9.33	6.93	5.95	5.41	5.06	4.82	4.64	4.50	4.39	4.30
	13	9.07	6.70	5.74	5.21	4.86	4.62	4.44	4.30	4.19	4.10
	14	8.86	6.51	5.56	5.04	4.70	4.46	4.28	4.14	4.03	3.94
	15	8.68	6.36	5.42	4.89	4.56	4.32	4.14	4.00	3.89	3.80
	16	8.53	6.23	5.29	4.77	4.44	4.20	4.03	3.89	3.78	3.69
	17	8.40	6.11	5.19	4.67	4.34	4.10	3.93	3.79	3.68	3.59
	18	8.29	6.01	5.09	4.58	4.25	4.01	3.84	3.71	3.60	3.51
	19	8.19	5.93	5.01	4.50	4.17	3.94	3.77	3.63	3.52	3.43
	20	8.10	5.85	4.94	4.43	4.10	3.87	3.70	3.56	3.46	3.37
	21	8.02	5.78	4.87	4.37	4.04	3.81	3.64	3.51	3.40	3.31
	22	7.95	5.72	4.82	4.31	3.99	3.76	3.59	3.45	3.35	3.26
	23	7.88	5.66	4.76	4.26	3.94	3.71	3.54	3.41	3.30	3.21
	24	7.82	5.61	4.72	4.22	3.90	3.67	3.50	3.36	3.26	3.17
	25	7.77	5.57	4.68	4.18	3.86	3.63	3.46	3.32	3.22	3.13
	30	7.56	5.39	4.51	4.02	3.70	3.47	3.30	3.17	3.07	2.98
	40	7.31	5.18	4.31	3.83	3.51	3.29	3.12	2.99	2.89	2.80
	60	7.08	4.98	4.13	3.65	3.34	3.12	2.95	2.82	2.72	2.63
	120	6.85	4.79	3.95	3.48	3.17	2.96	2.79	2.66	2.56	2.47
	∞	6.63	4.61	3.78	3.32	3.02	2.80	2.64	2.51	2.41	2.32

F-DISTRIBUTION 1% significance level

		\multicolumn{9}{c}{*Degrees of freedom for numerator*}
		12	15	20	24	30	40	60	120	∞
	1	6106	6157	6209	6235	6261	6287	6313	6339	6366
	2	99.4	99.4	99.4	99.5	99.5	99.5	99.5	99.5	99.5
	3	27.1	26.9	26.7	26.6	26.5	26.4	26.3	26.2	26.1
	4	14.4	14.2	14.0	13.9	13.8	13.7	13.7	13.6	13.5
	5	9.89	9.72	9.55	9.47	9.38	9.29	9.20	9.11	9.02
	6	7.72	7.56	7.40	7.31	7.23	7.14	7.06	6.97	6.88
	7	6.47	6.31	6.16	6.07	5.99	5.91	5.82	5.74	5.65
	8	5.67	5.52	5.36	5.28	5.20	5.12	5.03	4.95	4.86
	9	5.11	4.96	4.81	4.73	4.65	4.57	4.48	4.40	4.31
	10	4.71	4.56	4.41	4.33	4.25	4.17	4.08	4.00	3.91
Degrees of freedom for denominator	11	4.40	4.25	4.10	4.02	3.94	3.86	3.78	3.69	3.60
	12	4.16	4.01	3.86	3.78	3.70	3.62	3.54	3.45	3.36
	13	3.96	3.82	3.66	3.59	3.51	3.43	3.34	3.25	3.17
	14	3.80	3.66	3.51	3.43	3.35	3.27	3.18	3.09	3.00
	15	3.67	3.52	3.37	3.29	3.21	3.13	3.05	2.96	2.87
	16	3.55	3.41	3.26	3.18	3.10	3.02	2.93	2.84	2.75
	17	3.46	3.31	3.16	3.08	3.00	2.92	2.83	2.75	2.65
	18	3.37	3.23	3.08	3.00	2.92	2.84	2.75	2.66	2.57
	19	3.30	3.15	3.00	2.92	2.84	2.76	2.67	2.58	2.49
	20	3.23	3.09	2.94	2.86	2.78	2.69	2.61	2.52	2.42
	21	3.17	3.03	2.88	2.80	2.72	2.64	2.55	2.46	2.36
	22	3.12	2.98	2.83	2.75	2.67	2.58	2.50	2.40	2.31
	23	3.07	2.93	2.78	2.70	2.62	2.54	2.45	2.35	2.26
	24	3.03	2.89	2.74	2.66	2.58	2.49	2.40	2.31	2.21
	25	2.99	2.85	2.70	2.62	2.53	2.45	2.36	2.27	2.17
	30	2.84	2.70	2.55	2.47	2.39	2.30	2.21	2.11	2.01
	40	2.66	2.52	2.37	2.29	2.20	2.11	2.02	1.92	1.80
	60	2.50	2.35	2.20	2.12	2.03	1.94	1.84	1.73	1.60
	120	2.34	2.19	2.03	1.95	1.86	1.76	1.66	1.53	1.38
	∞	2.18	2.04	1.88	1.79	1.70	1.59	1.47	1.32	1.00

F-DISTRIBUTION   5% significance level

		\multicolumn{10}{c}{Degrees of freedom for numerator}
		1	2	3	4	5	6	7	8	9	10
Degrees of freedom for denominator	1	161	200	216	225	230	234	237	239	241	242
	2	18.5	19.0	19.2	19.2	19.3	19.3	19.4	19.4	19.4	19.4
	3	10.1	9.55	9.28	9.12	9.01	8.94	8.89	8.85	8.81	8.79
	4	7.71	6.94	6.59	6.39	6.26	6.16	6.09	6.04	6.00	5.96
	5	6.61	5.79	5.41	5.19	5.05	4.95	4.88	4.82	4.77	4.74
	6	5.99	5.14	4.76	4.53	4.39	4.28	4.21	4.15	4.10	4.06
	7	5.59	4.74	4.35	4.12	3.97	3.87	3.79	3.73	3.68	3.64
	8	5.32	4.46	4.07	3.84	3.69	3.58	3.50	3.44	3.39	3.35
	9	5.12	4.26	3.86	3.63	3.48	3.37	3.29	3.23	3.18	3.14
	10	4.96	4.10	3.71	3.48	3.33	3.22	3.14	3.07	3.02	2.98
	11	4.84	3.98	3.59	3.36	3.20	3.09	3.01	2.95	2.90	2.85
	12	4.75	3.89	3.49	3.26	3.11	3.00	2.91	2.85	2.80	2.75
	13	4.67	3.81	3.41	3.18	3.03	2.92	2.83	2.77	2.71	2.67
	14	4.60	3.74	3.34	3.11	2.96	2.85	2.76	2.70	2.65	2.60
	15	4.54	3.68	3.29	3.06	2.90	2.79	2.71	2.64	2.59	2.54
	16	4.49	3.63	3.24	3.01	2.85	2.74	2.66	2.59	2.54	2.49
	17	4.45	3.59	3.20	2.96	2.81	2.70	2.61	2.55	2.49	2.45
	18	4.41	3.55	3.16	2.93	2.77	2.66	2.58	2.51	2.46	2.41
	19	4.38	3.52	3.13	2.90	2.74	2.63	2.54	2.48	2.42	2.38
	20	4.35	3.49	3.10	2.87	2.71	2.60	2.51	2.45	2.39	2.35
	21	4.32	3.47	3.07	2.84	2.68	2.57	2.49	2.42	2.37	2.32
	22	4.30	3.44	3.05	2.82	2.66	2.55	2.46	2.40	2.34	2.30
	23	4.28	3.42	3.03	2.80	2.64	2.53	2.44	2.37	2.32	2.27
	24	4.26	3.40	3.01	2.78	2.62	2.51	2.42	2.36	2.30	2.25
	25	4.24	3.39	2.99	2.76	2.60	2.49	2.40	2.34	2.28	2.24
	30	4.17	3.32	2.92	2.69	2.53	2.42	2.33	2.27	2.21	2.16
	40	4.08	3.23	2.84	2.61	2.45	2.34	2.25	2.18	2.12	2.08
	60	4.00	3.15	2.76	2.53	2.37	2.25	2.17	2.10	2.04	1.99
	120	3.92	3.07	2.68	2.45	2.29	2.18	2.09	2.02	1.96	1.91
	∞	3.84	3.00	2.60	2.37	2.21	2.10	2.01	1.94	1.88	1.83

F-DISTRIBUTION   5% significance level

		\multicolumn{9}{c}{*Degrees of freedom for numerator*}
		12	15	20	24	30	40	60	120	∞
	1	244	246	248	249	250	251	252	253	254
	2	19.4	19.4	19.4	19.5	19.5	19.5	19.5	19.5	19.5
	3	8.74	8.70	8.66	8.64	8.62	8.59	8.57	8.55	8.53
	4	5.91	5.86	5.80	5.77	5.75	5.72	5.69	5.66	5.63
	5	4.68	4.62	4.56	4.53	4.50	4.46	4.43	4.40	4.37
	6	4.00	3.94	3.87	3.84	3.81	3.77	3.74	3.70	3.67
	7	3.57	3.51	3.44	3.41	3.38	3.34	3.30	3.27	3.23
	8	3.28	3.22	3.15	3.12	3.08	3.04	3.01	2.97	2.93
	9	3.07	3.01	2.94	2.90	2.86	2.83	2.79	2.75	2.71
	10	2.91	2.85	2.77	2.74	2.70	2.66	2.62	2.58	2.54
Degrees of freedom for denominator	11	2.79	2.72	2.65	2.61	2.57	2.53	2.49	2.45	2.40
	12	2.69	2.62	2.54	2.51	2.47	2.43	2.38	2.34	2.30
	13	2.60	2.53	2.46	2.42	2.38	2.34	2.30	2.25	2.21
	14	2.53	2.46	2.39	2.35	2.31	2.27	2.22	2.18	2.13
	15	2.48	2.40	2.33	2.29	2.25	2.20	2.16	2.11	2.07
	16	2.42	2.35	2.28	2.24	2.19	2.15	2.11	2.06	2.01
	17	2.38	2.31	2.23	2.19	2.15	2.10	2.06	2.01	1.96
	18	2.34	2.27	2.19	2.15	2.11	2.06	2.02	1.97	1.92
	19	2.31	2.23	2.16	2.11	2.07	2.03	1.98	1.93	1.88
	20	2.28	2.20	2.12	2.08	2.04	1.99	1.95	1.90	1.84
	21	2.25	2.18	2.10	2.05	2.01	1.96	1.92	1.87	1.81
	22	2.23	2.15	2.07	2.03	1.98	1.94	1.89	1.84	1.78
	23	2.20	2.13	2.05	2.01	1.96	1.91	1.86	1.81	1.76
	24	2.18	2.11	2.03	1.98	1.94	1.89	1.84	1.79	1.73
	25	2.16	2.09	2.01	1.96	1.92	1.87	1.82	1.77	1.71
	30	2.09	2.01	1.93	1.89	1.84	1.79	1.74	1.68	1.62
	40	2.00	1.92	1.84	1.79	1.74	1.69	1.64	1.58	1.51
	60	1.92	1.84	1.75	1.70	1.65	1.59	1.53	1.47	1.39
	120	1.83	1.75	1.66	1.61	1.55	1.50	1.43	1.35	1.25
	∞	1.75	1.67	1.57	1.52	1.46	1.39	1.32	1.22	1.00